Textbook on
Fish Food Organisms

The Author

Dr S.Athithan, who is currently as Professor & Head in Mandapam Centre for Sustainable Aquaculture, Ramanathapuram, is an accomplished teacher - researcher - extension worker - manager with a total of 23 years of work experience in various spheres of fisheries in Tamil Nadu. He received his Bachelor's Degree in Fisheries Science from Tamil Nadu Agricultural University (TNAU), Coimbatore in 1987 & Masters Degree in Fisheries Science from Tamil Nadu Veterinary & Animal Sciences University (TANUVAS), Chennai in 1990. He did Ph.D (Aquaculture) from TANUVAS, Chennai in 2004. He published a book on Composite Fish Culture Technologies in Tamil. He visited USA under Visiting Scholar Programme for three months during 2011 and learnt recirculatory aquaculture systems. He taught at both the undergraduate and graduate level. Besides served as resource person for various training programmes, organized 65 trainings & Co-Editor in KVK Newsletter (TANUVAS KVK), he has 32 peer reviewed research articles, 6 abstracts, 65 popular articles, 17 teaching / extension manual, 6 extension pamphlets and 2 text books to his credits. He has also given 10 radio talks in AIR, 6 TV programmes in Doordarshan. He documented 4 DVD on success story of fish farmers of Tamil Nadu. He also received various awards such as Distinguished Scientist Award in the field of Fish Nutrition, Best Professor Award in the field of Coastal Aquaculture, Best Researcher Award by Education Expo TV's and Best Innovative Extension Worker Award (2017). The current areas of Interest are quality control of aqua feeds, seaweed farming, recirculatory aquaculture systems, lobster fattening, integrated farming, acclimation studies of euryhaline marine fin & shell fishes to freshwater.

Textbook on Fish Food Organisms

S. Athithan

2026

Daya Publishing House®

A Division of

Astral International Pvt. Ltd.

New Delhi – 110 002

ISBN 978-93-89605-29-7

Published by : **Daya Publishing House®**
A Division of
Astral International Pvt. Ltd.
– ISO 9001:2015 Certified Company –
4736/23, Ansari Road, Darya Ganj
New Delhi-110 002
Ph. 011-43549197, 23278134
E-mail: info@astralint.com
Website: www.astralint.com

Digitally Printed at : **Replika Press Pvt. Ltd.**

Textbook on Fish Food Organisms

S. Athithan

2026

Daya Publishing House®

A Division of

Astral International Pvt. Ltd.

New Delhi – 110 002

ISBN 978-93-89605-29-7

Published by : **Daya Publishing House®**
A Division of
Astral International Pvt. Ltd.
– ISO 9001:2015 Certified Company –
4736/23, Ansari Road, Darya Ganj
New Delhi-110 002
Ph. 011-43549197, 23278134
E-mail: info@astralint.com
Website: www.astralint.com

Digitally Printed at : **Replika Press Pvt. Ltd.**

Preface

Aquaculture is now recognized as a viable and profitable enterprise worldwide. The past two decades have witnessed a dramatic expansion in the culture of marine finfish and crustaceans. Marine larviculture without live feed or crustacean cultures microalgae are rarities in commercial aquaculture. The development of commercial formulated feeds remains today's upcoming challenge. In the mean time, the industry continues the struggle to produce stable quantities of high quality live feeds. Cultivation of larvae is mostly done under controlled hatchery conditions and requires specific culture techniques different from the nursery and on growing procedures, especially with respect to zootechnics, feeding strategies and microbial control. The reason being that larvae are very small, extremely fragile and not fully developed compared to on growing juveniles. Therefore, larviculture nutrition and more particularly start feeding in the early larval stages, appears to be a major bottle-neck for the industrial up scaling of the culture of many fish and shellfish species. Therefore, this theory book is not written exclusively only for scientists but also for students who are practicing nutritionists and aquaculturists. The first part of teaching manual covers on Introduction, Types of feeds used in aquaculture, Importance of Natural food and Plankton in Aquaculture, Plankton sample collection and protocols, Candidate species of Live feed organisms, Current Status Of Live Feed Production For Larviculture and Nutritional Properties Of Microalgae. The second part includes Types of algal cultures Algal Culture Techniques Methods of Isolation of Micro algae, Physical and chemical conditions for Algal Culture, Growth dynamics of Algal Culture and Algal Culture media. The last part of this manual is dealing with Biology and Culture Requirements of Rotifers, Artemia, Spirulina Cultivation, Earth Worm Cultivation, Culture of Infusorians, Culture of Bloodworms, Culture of Tubifex, Culture of Moina, Culture of Daphnia and Bioenrichment of Artemia nauplii. Each and every part has been meticulously prepared and fortified with new information. Nonetheless, the material presented has been thoughtfully selected and updated to make it of maximum use to persons whose interests range from general aquaculture to animal nutrition to live feed manufacture and student communities in particular.

S. Athithan

Contents

Contents

Chapter 1

Introduction

Based on the dietary needs of the fish body (nutrient requirement), physiology of digestion and food & feeding habits of fish, one can develop a feed conforming scientific formulation of feed, optimal processed feed and supply of feed judiciously to fish. Cultivating organisms such as Plankton (phyto & zooplankton), Annelids, Molluscs, Amphibians, Bait & Forage fishes that form food either natural or substitute for fish, which are collectively called as "Fish Food Organisms".

Food and Feeding Habits of Fish

Diverse habitats of fish include 41.2 % in freshwater, 58.2 % in marine and 0.6 % an diadromous, migrating between fresh and saline waters. Based on the nature of food and habitats of fish, wide range of feeding habits of fish includes the following 3 primary groups

Herbivorous Fishes: They feed on plant material, which forms more than 75 % of gut contents. Herbivores are divided into 2 sub groups.

- Planktophagous fishes: They consume only phyto and zooplankton
- Detritophagous fishes: They feed on detritus.

Omnivorous Fishes: They consume both plant and animal food. Omnivores can also be grouped into 2 categories.

- Herbi-omnivores: Fishes feed more on plant material than animal food,
- Cami-omnivores: Fishes feed more on animal food than plant material.

Carnivorous Fishes: They feed on animal food, which comprises of more than 80% of the diet. Carnivores are also classified into

- Insectivores (feed on insects)
- Carcinophagous (feed on crustaceans)
- Malacophagous (feed on molluscans)

- Piscivorous (feed on other fishes) and Larvivorous (feed on larvae).

Some fishes are cannibalistic. The fishes consume a variety of food material, such as phytoplankton, zooplankton, aquatic weeds, animals like annelids, arthropods, molluscs, other fishes and amphibians.

Selection of Food

When selecting food to be fed to the larvae, the following points should be considered:

1. The food must be perceived by the larvae. From this point of view, live food is preferred to inanimate food during the first few days of feeding.
2. The size of food must be such that it can be accommodated by mouth of larvae.
3. The feed should have high dietary value especially Highly Unsaturated Fatty Acids (HUFA) essential to the growth and survival of the larvae.
4. The feed can be easily produced in large quantities.
5. The feed can be digested by the larvae.

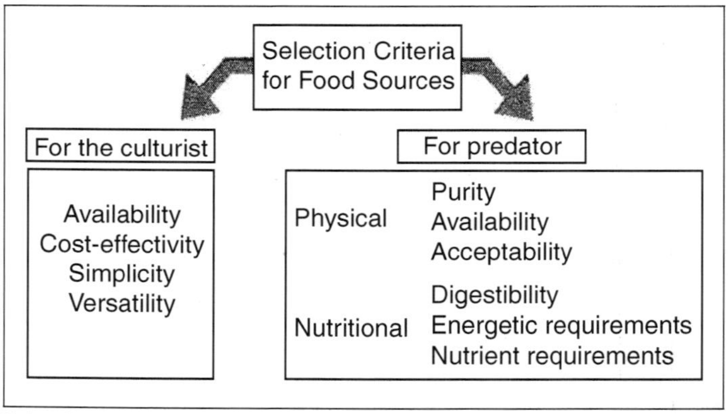

Selection Criteria for Larval Food Sources from Culturist Side and the Cultured Larvae

Chapter 2

Types of Feeds Used in Aquaculture

Successful production good quality seed and table size fish can be achieved, by and large, through the operation of 2 regimes, viz, feeding the larvae and fish with nutritionally balanced feed; maintain the water quality parameters in optimum range. The feed is differentiated by its physical properties and its contents. There are many varieties of feed used in aquaculture system. The feed used by the farmer is according to their convenience. The wrong usage of feed must be affecting the aquaculture system. The feed is broadly classified into two types such as natural and artificial feed.

Natural Feed

It is otherwise called as viable feed or live feed. Natural feeds generally serve as living capsules of nutrition. They are live feed or naturally occurring substances which can perform the growth prominently. These feeds are very cheap and occur very easily. Live feeds do not have most of the disadvantages of artificial feeds.

Natural feeds are further classified as aquatic and non-aquatic live feeds. The examples for aquatic live feeds are

Plankton

Fish production in a water body is directly or indirectly dependant on the abundance of plankton. The physico-chemical properties of water determine the quality and quantity of plankton. Thus, during the study of plankton, a link in the food chain is a pre-requisite to understand the capacity of the water body to support the fisheries and the need for introduction of additional selected species of commercially important fishes. Other two categories of life in an ecosystem are benthos and nekton. Benthos is the term given to life at the bottom, like aquatic earthworms, insect larvae and certain fishes. Nekton includes the larger swimming animals like fishes. Plankton is most essential for many fishes as food. The growth of plankton feeding fishes mostly depends on plankton dynamics of the water body.

The plankton is further divided into two main categories namely phytoplankton and zooplankton.

Phytoplankton (Unicellular Algae or Filamentous Algae)

Fishes consume the phytoplankton, which is found abundantly in well managed ponds. Phytoplankton gives green colour to the water due to the presence of chlorophyll. Phytoplanktons are generally made up of mostly unicellular algae which are either solitary or colonial. Phytoplanktons are autotrophs, i.e., they fix solar energy by photosynthesis using CO_2, nutrients and water. Phytoplankton occupies the base of the food chain and produces the food material on which other organisms in the ecosystem sustain themselves. The phytoplankton drifts about at the mercy of the wind and water movements. Algae consist of three major classes which form the main food in phytoplankton. These are chlorophyceae, cyanophyceae and bacillariophyceae.

Chlorophyceae

These are called green algae due to presence of chlorophyll. Many chlorophyceae members are useful as food to fishes. These organisms are distributed all over the pond. The chlorophyceae members useful as fish food are *Chlamydomonas. Volvox, Eudorina, Pandorina, Chlorella,* filamentous algae like *Ulothrix, Oedogonium, Spirogyra, Pediastrum, Microspora, Cladophora, Clostredium, Scenedesmus, Cosmarium.*

Cyanophyceae

These are also called as myxophyceae and are commonly known as blue green algae. This colour is due to the varying proportions of chlorophyll a, carotenoids and biliproteins. The product of photosynthesis is cyanophycean starch, present in granular form. The cell wall lacks cellulose and instead comprises mainly of amino acids and amino sugars. Many cyanophycean members are consumed by fishes. These are *Nostoc, Oscillotoria, Anabaena, Microcystis, Spirulma, Merismopedia, Arthrospira* etc.

Bacillariophyceae

These are called diatoms. They are unicellular organisms with different shapes and sizes. These may be yellow or golden brown or olive green in colour. Golden brown diatomin pigment is present in diatoms. The reserve food materials are fat or volutin. The diatoms consumed by fish are *Diatoma, Navicula, Cocconies, Synedra, Tabellaria, Meridian, Fragilaria, Nitzschia, Pleurosigma, Amphifileura, Rhizosolenia, Cyclotella, Amphora, Melosira, Aclwanthes* etc.

Zooplankton

Plankton consisting of animals is called zooplankton. Zooplankton is abundant in the shallow areas of a water body. The zooplankton unlike phytoplankton is particularly distributed horizontally and vertically in an ecosystem. They also undergo diurnal vertical migrations. The zooplankton forms an important group as it occupies an intermediate position in the food web, many of them feeding on algae and bacteria and in turn being fed upon by fishes. They also indicate the trophic status of a water body. Their abundance increases in eutrophic waters. They are also sensitive to pollution and many species are recognized as indicators of pollution. Among zooplankton, some of the organisms occasionally occur in appreciable numbers forming swarms. These swarms occur in freshwater ponds forming bands or streaks, or are arranged into areas of thick and thin concentration. Simulating cloud effect, they may give the water a strikingly different colour in the region of the swarm. The most common organisms in zooplankton are protozoans, crustaceans and rotifiers

Protozoa

Protozoans are most primitive, unicellular and microscopic animals. These organisms are provided with locomotory organellae like pseudopodia, flagella and cilia. These organisms are found abundantly in fish ponds and are useful as natural fish food. Generally, protozoans dominate in the zooplankton communities. Protozoans in general, are solitary single celled organisms. In some dinoflagellates and ciliates, the daughter individuals do not separate and together form pseudocolonies during repeated fission. These colonies are called catenoid colonies. The protozoans belonging to the classes sarcodina, flagellate and ciliata are useful as food items to fishes. The protozoans with pseudopodia are included in the class sacrodina or rhizopoda. *Amoeba* and *Actinophrys* are common sarcodines found in fish ponds and are also used as food by fishes and prawns occationally. The protozoans with flagella are included under the class flagellata or mastigophora. *Euglena* is the most common fish food organism found in fish and prawn ponds. *E. viridis, E. spirogyra* and *E.minuta* are some important species used as fish food. *Ceratium, Chilomonas* and *Phacus* are also used as fish food. The protozoans with cilia are included in the class Ciliata. Here, the cilia persist throughout life. The ciliates like *Paramecium, Vorticella. Coleps, Colpoda. Metropus, Euplotes. Oxytricha* etc. are fish food organisms. The ciliates are the dominating organisms among the zooplankton.

Crustacea

The aquatic animals with 19 pairs of appendages and branchial respiration are included in the class Crustacea. The crustaceans vary form microscopic to large animals. Crustaceans form a major component of zooplankton. In zooplankton, the micro crustaceans are useful as food to fish and prawns. The important micro crustaceans are copepodes and cladocerans. The crustacean nauplii also constitute good food material for many fishes and prawns. For example, nauplei *of Artemia* are used in prawn hatcheries.

a) Copepoda

These are animals with 5 pairs of thoracic appendages, abdomen without appendages, forked telson, two pairs of antennae and body with head, thorax, and abdomen. The copepodes inhabit many of the freshwater habitats such as lakes, reservoirs, ponds, etc. Many of the copepodes are pelagic and so are abundant in the plankton of both limnetic as well as littoral regions of the water. Only the harpacticoids are mostly benthic or bottom living. The size of the body of the copepods is 0.3 to 3.5 mm. Copepods such as *Cyclops, Mesocyclops, Diaptomus, Canthocamptus,* etc. are useful as fish food organisms.

b) Cladocera

The animals which are bivalved, shield shaped with or without shell, flattened trunk appendages and leaf like caudal styles which may be unjointed or jointed are included in cladocera. The greatest abundance of cladocerans is found near the vegetation in lakes, ponds, etc. The size of these shelled crustaceans varies from 0.2 to 3.0 mm. The cladocerans like *Daphnia, Ceriodaphnia, Moina, Sinocephalus, Scapholebris, Sida, Eurycents, Chydorus, Daphniosoma, Polyphemus, Macroihrix, Leydigia etc.* are useful as fish food organisms.

c) Ostracoda

The animals with bivalved carapace, which encloses the entire body, 4 - 6 trunk appendages and reduced trunk are included in ostracoda. These forms are well represented in both the standing and running waters. These are exclusively planktonic forms. Occasionally the ostracods like *Cypris, Stenocypris,* etc are consumed by fish.

d) Rotifera

Rotifers are readily identifiable from other planktonic materials by the presence of their anterior ciliated wheel-like structure called corona and hence they are called wheel animalcules. Rotifers live in a variety of aquatic habitats. They are microscopic, ranging from 40 microns to 2.5 mm in size. Usually rotifers like *Keratella, Phlodina, Rotaria, Hexanhra, Filinia, Brachionus Epiphanes* etc are useful as food organisms. Rotifers offer several advantages as fish feed organisms. They are

They reproduce quickly, it is estimated that a population under favourable conditions can double every one to five days. Under intensive laboratory conditions, they have recently been reported to have a doubling rate of less than 9 hours.

Rotifers are small and therefore are accepted as food by some organisms that cannot ingest larger zooplankton. Thus, they are an important first food for many fishes and prawns. They are nutritious and their actual nutritional value can be improved, as can be done for other zooplankton, by packing the rotifers with specific strains of algae or other feed.

Aquatic Weeds

Though the aquatic weeds form undesirable vegetation, which cause damage to the fisheries, these are helpful as food for a few fishes. Many herbivorous fishes consume aquatic weeds. The grass carp is a fast growing fish that feeds on aquatic weeds. This fish utilizes submerged weeds like *Hydrilla, Najas, Ceratophyllum, Ottelia, Nechamandra and Vallisneria.* The young fish prefer smaller floating plants like *Wolffla, Lemna, Azolla* and *Spirodela* and the other herbivorous fish utilize aquatic weeds are *Pulchellus pulchellus, Tilapia* and *Etroplus.* Though an omnivore, common carp feeds well on filamentous algae like *Pithophora* and *Cladophora. Trichechus sp,* a large air-breathing herbivore, is being utilized for clearance of aquatic weeds in the canals of Guyana.

Annelids

Animals with metameric segmentation, eucoel, nephridia and setae are included in the phylum annelida. The animals which belong to classes polychaeta and oligochaeta are useful as fish food organisms. These are found at the bottom of the water body and are generally consumed by bottom-dwelling fish, common carp, catfishes, murrels, etc. *Tubiflex, Glycera* and earthworms are the common fish food oligochaetes.

Insecta

Animals with 3 pairs of legs, 2pairs of wings, jointed appendages and a chitinous body wall are included in class insecta. Insects and their larvae form the main food item of many fishes. Aquatic insects are often preyed upon by fish like trout, catfishes, murrels etc. Hemiptera, diptera, coleoptera, ephemeroptera and plecoptera insects dominate as fish food among the insects. Belostomatidae & notonectidae and nymphs of odonata are good fish food organisms. Larvae of mayflies, dragonflies, chironomid larvae, chaoborus larvae and mosquito larvae are also found commonly in fish diets. When mayflies constitute the diet of trouts, it has been observed that the trout are fatter and better flavoured.

Mollusca

The animals with a soft body, shell and foot are included in the phylum mollusca. The molluscans are found at the bottom of water body. Hence, only bottom dwelling fish consume them. The gastropods are found in the diets of carnivorous and omnivorous fishes. For some marine fish species (i.e. siganids, groupers, snappers) very small zooplankton, such as trochophora larvae need to be used as a starter feed, since the commonly used rotifers are too big. Trochophora larvae of the Pacific oyster *Crassostrea gigas* are 50 µm in size and free-swimming (slow circular swimming pattern) ciliated organisms which have a high nutritional value for marine fish larvae. For example, trochophora larvae may contain up to 15 % (of total fatty acid) of both EPA and DHA.

Amphibia

Amphibians are tetrapodes, terrestrial as well as aquatic. The fishes consume only anuran larvae, the tadpoles among amphibians. The consumption of tadpole larvae is not frequently found.

Nematodes

The use of the free living nematode, *Panagrellus redivivus* as larval food has been demonstrated successfully for several species, including *Crangon crangon*, juvenile king shrimp (*Penaeus blebejus*), common carp (*Cyprinus carpio*) and silver carp (*Hypophthalmichthys molitrix*).

Forage or Bait Fishes

Carnivorous (piscivorous) fishes feed on a variety of other adult fishes, fish eggs, fry and fingerlings. Fishes like murrels, freshwater shark, seenghala, etc. feed on other fishes. The small fishes like *Salmostoma, Amblypharyngodon, Puntius, Labeo, Chanda, Nuria, Lebistis, Gambusia, Esomus* etc. are consumed as food by larger fishes. Some fishes are cannibalistic in nature. Fishes also feed on decapods (prawns). The carnivorous and omnivorous fishes feed on small prawns. For example, *Macrobrachium kitsuensis* is found in the gut of many fishes. *Acetus* prawn suspension is given as food to the larvae and post larvae of prawns in the hatcheries. The bryozoans or ectoprocta are also found in the gut of fishes. They enter accidentally into the mouth of fishes. Generally bryozoans inhabit the aquatic weeds, stones and pebbles. When aquatic weed are taken as food by fish, along with the weeds the bryozoans enter into the mouth of the fish. Adult bryozoans and statoblasts *of Bugula* and *Hyalinella* are found in fish gut.

Summary

A variety of natural fish food organisms are found in a water body, which depend on the nutritive nature of the water body. The natural food provides the constituents of a complete and balanced diet. The natural fish food organisms are plankton, oligochaetes, insect's larvae, molluscs, tadpoles, weeds, etc. The plankton is divided into two main categories – phytoplankton and zooplankton. The phytoplankton drifts about at the mercy of the wind and water movements. Algae consist of three major classes which form the main food in phytoplankton. These are chlorophyceae, cyanophyceae and bacillariophyceae. The most common organisms in zooplankton are protozoans, crustaceans and rotifiers

Chapter 3

Importance of Natural Food and Plankton in Aquaculture

A variety of natural fish food organisms are found in a water body, which depend on the nutritive nature of the water body. The natural food provides the constituents of a complete and balanced diet. The demand of natural food varies from species to species and age group of individuals. For example, *Catla* prefers zooplankton and silver carp prefers phytoplankton. At a younger stage; the fish may feed on plankton, and the same fish may prefer animal food as an adult Fishes feed on different natural food organisms at all the different trophic levels. Natural feeds have high protein and fat contents, which promote the growth of the fish. Hence, it is necessary to increase the live food in the aquatic ecosystem to improve the growth of the fish. Whereas in the 1970s the production of farmed marine finfish and shrimp relied almost exclusively on the capture of wild fry for subsequent stocking and on-growing in ponds, tanks or cages, the complete domestication of many marine and brackishwater aquaculture species was only achieved during the last two decades. However, since then the controlled production of larvae from captive broodstock, or in other words the hatchery production of fry, has now become a routine operation for most cultivated fish and shellfish species; billions of fish and shellfish larvae (i.e. bivalve molluscs, penaeid shrimp, salmonids, European sea bass, Gilthead sea bream etc.) currently being produced within hatcheries all over the world. The following objectives are associated with the phytoplankton:

- To increase dissolved oxygen and to decrease toxic gases like ammonia, nitrite, hydrogen sulfide, methane, carbon dioxide in pond water
- To stabilize pond water quality and to lower the content of toxic compounds
- To make use of plankton as a natural feed
- To provide shade and to decrease cannibalism
- To increase and stabilize water temperature

- To minimize pathogenic and unwanted microbial population by competing with the available nutrients in the water

Golden Brown or Reddish Brown

This kind of water color has a high probability of comprising diatoms in the pond. It occurs frequently due to lower temperature, and in saline waters having high organic matter concentration. Transparency is normally between 25 and 35 cm. The blooming of diatoms causes this colour. Algae species such as *Chaetoceros, Navicula, Nirzschia, Skeleronema, Cyclotella, Synedia, Achnanthes, Amphora* and *Euglena* are often found in pond water of this colour, especially the first three species. This colour is quite difficult to achieve. Diatoms are abundant in nutrients and are easily digestible. Golden brown color is usually related to a crop of healthy shrimp, with brilliant body color and is an indicator of expected good yield.

Light or Bright Green

This colour is due to the growth of green algae, especially *Chlorella.* In addition, *Dunaliella, Plarymonas, Carteria, Chlamydomonas* are also present. In less saline water, *Scenedesmus* and *Euglena* can also be found. Water of this colour is usually quite stable. In other words, the mortality of prawns is low. Transparency is usually between 20 and 70 cm. It usually occurs in a new pond, sandy bottom pond and ponds with less organic sediments. The growth rate of shrimp / prawn / fish in this environment is quite stable, although it is comparatively slower than that in the brownish water and they grow with less variability of size. This kind of watercolor is easy to manage. Culture animals in a pond are susceptible to disease when the green water turns too dense with a dark green color. However, it seldom causes high mortality.

Dark Green or Blackish Green Colour

When the pond water temperature gets too high, deterioration of pond bottom or organic materials accumulate too fast, blue green algae bloom faster than green algae. Blue green algae such as *Oscillatoria, Phormidium* and *Microcoleus* dominate (accounting for 90 %). Although the survival rate of shrimp in ponds with this colour still remains high, the growth rate sinks. The appearance of shrimp in this watercolor becomes dark green or acquires a black colour. Sometimes it is even bad looking having been infested with protozoa and germs. In this environment, shrimp grow slower and unevenly with great difference between large one and small one and a high degree of skewness of size. Furthermore, it causes higher percentage of soft shelled shrimps and blue-sky coloured shrimps especially in low salinity conditions. It also weakens the shrimp. It is not a desirable color in semi-intensive culture and should be transformed and improved through management.

Dark Brown and Sauce-like Colour

Poor pond management such as overfeeding or using large amounts of trash fish causes a rapid growth of dinoflagellates and brown algae and consequently results in the formation of this water colour. Such water conditions are undesirable and it is recommended to change the pond water partially if this colour arises. It indicates super-eutrophication that is usually caused by over feeding, deterioration of pond bottom, low water exchange rate, high concentration of organic matter and high application of trash fish. Transparency of this water is about 15 cm. It forms a lot of bubbles while using aerators. It is not a desirable watercolor in pond management because it usually weakens the shrimp and sometimes causes dropsy and makes the animal susceptible to gill disease. It always produces a poor harvest. The reason for this dark brown colour to form is not due to algae, but by the pigment and tannic acid, that normally occurs a high concentration in mangrove area because of acid sulfate soils.

Yellowish Colour

The formation of yellow water is due to the growth of *Chrysophyta.* In addition, green flagellates may also grow moderately. In a pond with watercolour of this type growth of benthic diatom, blue green algae and green algae is inhibited. This kind of water is very stable in colour, although it is not a desirable one for the shrimp / prawn culture. It is usually formed in old fish ponds that have accumulated a high organic load over a long period and through bacterial action. In ponds with water of this colour, growth of shrimp / prawn is inhibited and chances of high mortality exist. Because all these algae / phytoplankton are very small in size, they cannot be used directly as natural foods by the shrimp / prawn. The phytoplankton forms an indirect source of food to shrimp/prawn through zooplankton as thy consume phytoplankton.

Foggy White Water

This mainly comprises of zooplankton, clay particles and detritus. It is an ideal environment for growth of fry or juvenile prawns since it provides natural feeds. On the other hand, for the adult prawn there exists a disadvantage of competing with the environment during the culture period. In waters of this colour, zooplankton gathers together and gives an appearance of being suspended as a white spot. Too high a population of zooplankton in the environment interferes with shrimp behaviour and can cause damage to shrimps. It is also indicative of the die-off of algae or the deterioration of water quality with resulting propagation of bacteria. This water colour will need to be well controlled during culture period.

Turbid Water

The formation of turbid water may be due to the suspension of zooplankton, clay particles or detritus. This kind of water can be beneficial or harmful, depending on the quality and quantity of the suspended materials. It is formed by clay particles that originate from the dike and are dissolved by rain or from the river where freshwater is sourced. Sometimes, it is caused by strong wind and the resulting disturbing movement of shrimp at the bottom. This turbid water can provide some advantages not only to stabilize the water quality and shrimp habitat, but also provide some nutritional effect since the clay particle can absorb nutrients, organic particles and microorganisms to form "clay floccules" which can serve as shrimp natural feed. However, if the original water is clear, unlike the water that had the colour as discussed earlier, too many clay particles can be negative to the growth of shrimp.

Clear Water

This water is transparent. This may be caused by a lack of nutrients, the presence of heavy metal pollutants like copper, manganese, iron or acid bottom clay (pH 5.5 or lower). Under these conditions, no organisms can grow properly. It goes without saying that this kind of water cannot raise a good crop of shrimp / prawn / fish, because it indicates that something is wrong with the soil or the water quality. This must be improved at the start of shrimp / prawn / fish culture. It is evident that some types of water colour are desirable, some are not. In order to achieve a particular colour, algal nutrients may be used. It is suggested that ammonium salts are good for green algae growth, while urea is good for brown algae (i.e. diatoms). When the colour becomes undesirable owing to over blooming, bactericides, insecticides and algaecides may be used. Increasing aeration and / or partial replacement with clean water may also be helpful in changing water quality. Feeding greatly influences water colour and water quality. Over feeding should be avoided. The use of too much trash fish may cause a blooming of flagellates, which is not desirable.

Transparency

There is a high correlation between secchi disc visibility and phytoplankton (water colour) density. Secchi disc visibility is the average of the depth at which a disc, a round plate with alternating black and white quadrants disappears and reappears from view when sunlight is intense or moderate. The optimum range for secchi disc reading is between 30 and 60 cm to the juvenile stage and between 25 and 40 cm to the sub adult and final stage. High secchi disc reading is associated with low productivity of the pond and low secchi disc reading is associated with high productivity of the pond. In general, high productivity ponds are having high biomass that increases oxygen consumption, which may lead to oxygen depletion.

Secchi Disc Measurements

Secchi disc is a simple circular plate / disc with alternate segments painted black and white. The plate/disc is mounted at the end of a rod / stick upon which distances are marked with zero being at the plate end. To take transparency measurements a farmer can either stand on walkway platform or at the edge of the dike and lower the secchi disc holding the end of the stick slowly into the water. The depth at which the black and white segments of the disc disappear is the secchi disc reading.

Zooplankton

To feed the delicate larvae, juveniles and even adults of fish, prawn and shrimp in nurseries and grow-out ponds protein rich and naturally occurring zooplanktonic organisms are essential. Different types of zooplanktons like larval forms of brine shrimp (*Artemia salina*), rotifers (*Brachionus* sp.), cladocerans (*Moina* sp/ *Euchlanis* sp/ *Daphnia* sp/ *Ceriodaphnia* sp), copepods (calanoid, cycloid and harpacticoid copepods) and larval forms of different aquatic organisms etc. are considered as natural food for the prawn / shrimp / fish.

Availability in the Pond

- Take pond water in a glass and observe closely through. Greenish to brownish colour and the presence of actively moving small organisms is indicator of the availability of fish food organisms in the pond.

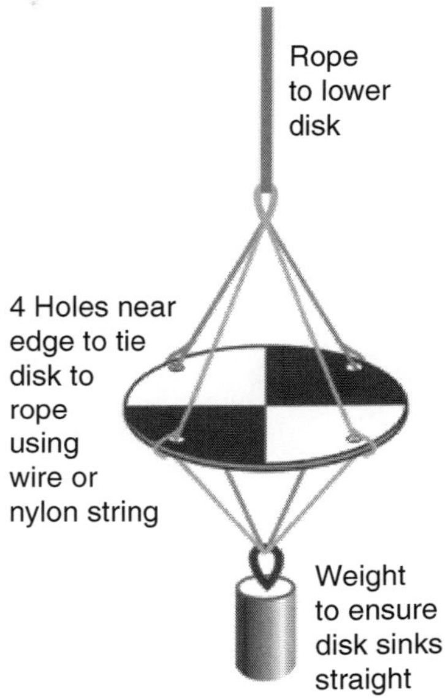

Chapter 4

Plankton Sample Collection and Protocols

The largest plant community in the oceans is the free drifting, microscopic phytoplankton. Phytoplanktons are unicellular, microscopic, unattached plants found in freshwater and marine ecosystems. Bi weekly to monthly sampling is necessary to capture the seasonal dynamics of phytoplankton and to quantify their abundance and biomass. Zooplankton can be collected from seawater bodies as well as freshwater lakes or ponds.

Microalgal cells can range in size from less than 1 mm to greater than 1 000 µm, and in some cases can multiply to more than one million cells per liter during a high peak bloom. The phytoplankton evolved into a diverse group of photosynthetic organisms that are divided intocategories according to size:

- Microplankton (20 - 200 µm)
- Nanoplankton (10 - 20 µm)
- Ultraplankton (2 - 10 µm)
- Picoplankton less than 2 µm

Picoplankton and ultraplankton tend to be more common in the open ocean than the coastal waters, whereas the larger nanoplankton and microplankton, such as diatoms and dinoflagellates, tend to dominate coastal waters.

There are various methods for collecting samples for phytoplankton analysis, depending on whether a quantitative or qualitative analysis is desired. Irrespective of methods, samples should be preserved soon after collection and where possible, live samples should also be examined.

Qualitative Sampling

Plankton net sampling - involves lowering standard plankton net to just off bottom and towing it gently in a vertical direction to obtain an integrated sample throughout the water column. Plankton nets have the advantage of simultaneous collection of larger quantities of plankton, giving larger concentrations of the rarer cells and aiding in their identification. The best size for mesh spacing is 20 µm (sometimes 10 µm). The ratio of net length to net mouth diameter should be between 3:1 and 5:1.

In order to account for horizontal patchiness of algal species the plankton net may be towed in a horizontal direction at depth 0.5 – 1 m below the surface. This method does not allow for accurate counting of algal species. In addition, some phytoplankton species pass through the net, or burst or disintegrate when stressed or coming in contact with the net.

Quantitative Sampling

For quantitative phytoplankton sampling the most conventional procedure is sampling at discrete depths by sampling water bottles

Fixatives/ Preservatives

Preservatives are toxic by definition. Delicate organisms without strong cell walls may collapse before fixation is complete. To minimize this effect, it is important that the fixative agent enters the cell quickly, which may be achieved by the addition of dimethyl sulphoxide. Some preservatives, *e.g.*, acid solutions of Lugol may eventually lead to the loss of some groups of organisms. Preservatives have to meet the following requirements:

- The effect of the agent on the loss of organisms by chemical shock or otherwise must be known before.
- The preservative must effectively prevent the microbial degradation of organic matter at least during the storage period of the samples.
- The preservative must guarantee a good recognition of taxa at least during the storage period of the samples.

Collection Techniques

Harvesting techniques depend strongly on the location of the harvesting site and should meet the following criteria:

- Capable to operate on a continuous basis without surveillance
- Easy to transport and to set up
- Relatively cheap in purchase and maintenance
- Available on site
- Designed for the required quantities and zooplankton sizes

The following mesh sizes may be used to collect the various sizes of freshwater zooplankton:

- 80 µm for small species of rotifers and larger infusorians. These are an excellent

- starter feed especially for the fry of some fishes that need small food in the early stages (tench, grass carp, silver carp, big head, carp)
- 160 μm for larger rotifers, nauplius and copepodite stages of copepods
- 300 and 500 μm for small water fleas and smaller species of cyclopoid copepods
- 700 μm for adult water fleas of the *Daphnia* genus, large species of cyclopoid and calanoid copepods, larvae and pupae of *Corethra sp., etc.*

There are three main methods of zooplankton collection, which are as follows:

Bottles/ Water Samplers

This method is used mainly for collecting smaller forms or microzooplankton.

Pumps: The gear is normally used on board the vessel / boat.

Nets: The most common method of zooplankton collection is by a net. Soak the body of the zooplankton net in lake water prior to use (2 minutes). Rinse the net with lake water to dislodge any attached material prior to sampling.

The net is conical shaped, 3 - 3.5 m long, the inlet opening is 1 - 1.2 m in diameter and the end hole has a diameter of 0.2 - 0.5 m. There is a strip of thicker cloth on both ends; the front end is furnished with buoys to allow the net to be fixed to a frame. The rear end may consist of a PVC cylinder (2 l), which can be closed on one side. Nets for hand collection of zooplankton are of the sac type, 50 - 60 cm long. The net is fixed to a metallic ring, 40 - 50 cm in diameter, held on a rod of about 2 m long. Collecting zooplankton with hand nets is rather in efficient: one person can catch about 0.1 to 1.0 kg of plankton per hour, depending on the amount of zooplankton biomass in the reservoir.

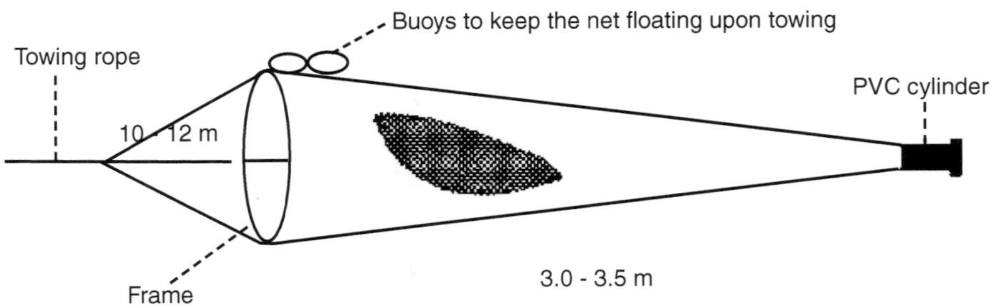

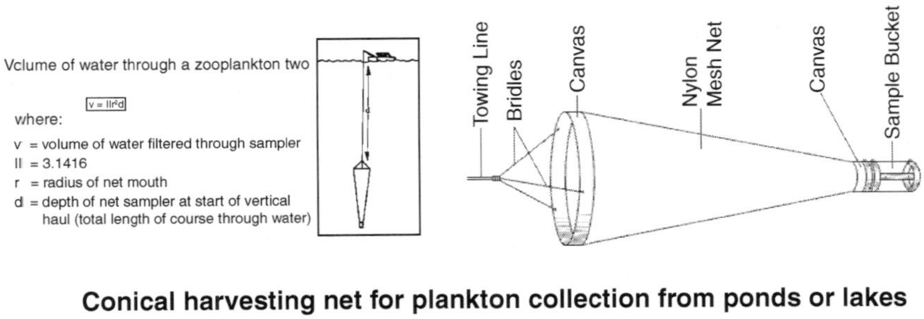

Conical harvesting net for plankton collection from ponds or lakes

Tube Method

The third method is a tube, made of common PVC or Tygon. A tube is lowered into the water column and when the bottom reaches the desired depth (near the sediments); a line is pulled to close the bottom with a rubber stopper or other device. The tube is raised into the boat and the collected water poured through a net to collect the zooplankton. This device also provides a high degree of certainty about volume of water sampled, but it may not be an effective way to sample large animals that occur at a low density, or animals that can detect and escape from a narrow sampling device.

Tubes can be regarded as a sort of long bottle, are usually flexible, and they collect an integrated sample when lowered through the water column. Their length must be equal to the maximum depth to be sampled. The internal cross section of tubes should be as large as possible in order to minimize the escape of zooplankters that are rapid swimmers. Tubes with a small diameter, however, can be closed more simply and do not require special closing mechanisms. Flexible tubes are easily managed from small boats, but they must be weighted at the immersed end so that they remain vertical. If the weight is made of metal rings which are spaced at 10 cm intervals from one another and are connected to a small rope the same length as the tube, they can constitute an efficient system for closing the tube. The tube finds its greatest application in sampling an entire column of water in shallow environments or in littoral areas rich in vegetation. Transparent rigid tubes can also be employed for sample collection in plastic bag enclosures where a representative sample of a small volume is needed, or in shallow littoral waters, ponds, and pools. One should remember that integrated tube samplers weight all strata sampled as if they contribute equally to the lake volume. In addition, all information, regarding the depth distribution of zooplankton is lost. This can introduce important errors to production calculations.

Plankton Light Trapping

A more elegant method for zooplankton collection takes advantage of the positive phototactic behaviour of some zooplankton species. The effectiveness of light to attract the zooplankters is directly dependent on the water transparency and on the intensity of the light source. It is useless to apply this method where the water transparency is below 30 cm. Cladoceran and cyclopoid copepods respond most sensitively to light, rotifers less. The best results of collecting zooplankton with light are obtained in the early night (until about 10 pm); later the effectiveness declines. Though the success of this method may vary, the low expenditure necessary for its application seems to make it an economically viable harvesting system for freshwater species.

Procedure

1. Soak the body of the zooplankton net in lake water prior to use (2 minutes)
2. Rinse the net with lake water to dislodge any attached material prior to sampling

3. Attach the zooplankton bucket - make sure the plug is in place
4. Fill the bottle with lake water that has been filtered through the net mesh
5. Lower net to the euphotic zone depth making sure it stays in a vertical position
6. Raise the net vertically at a continuous rate of 0.5 m/ sec to minimize avoidance of the net by fast-swimming zooplankton
7. Avoid sampling near sediments & macrophytes because non-planktonic species of Rotifers & Crustaceans inhabit these substrates & would contaminate plankton
8. At surface, rinse down outer sides of net two to three times with lake water. Do not splash rinse water into the net opening, or let it drop below the surface
9. Separate the bucket from the net, place the lower end of the bucket into an open sample jar, then remove plug & drain the zooplankton & water into jar
10. Rinse the bucket contents into the sample jar with squeeze bottle previously filled with filtered net water
11. Zooplankton samples can be preserved in either 95 % ethanol or 5 % formaldehyde. Formaldehyde is preferred because counting samples preserved in ethanol is difficult because of the convection currents caused by rapid evaporative losses. The following approaches can be used to reduce distortion due to Formalin preservation: (a) addition of 40 g/ L of sucrose to formaldehyde solutions; (b) maintenance of samples at low temperature (6°C); and (c) narcotisation with carbonated water or methanol prior to preservation in a formaldehyde-sugar solution (Paterson, 2003).
12. Rinse the net and bucket with lake water between sites.
13. Record sampling location (GPS coordinates), site, date, time, samplers ID, number of hauls and depth of haul on the jar and field sheets/book. Also note the mesh size and dimensions, the fixative used and the prevailing weather conditions in the field sheets / book. Note and record ice depth in the winter.
14. Put a few drops of glycerin into sample when back at lab, before storage of sample. This helps prevent the animals from sticking together.

Chapter 5

Candidate Species of Fish Food Organisms

Phycology (from Greek word, *phykos*, "seaweed"; and logy, studies) is the scientific study of algae. Phycology is a branch of life science and often is regarded as a sub discipline of botany. Phycology includes the study of prokaryotic forms known as blue-green algae or cyanobacteria. A *phycologist* is a person who studies algae as described above.

Algae are important plants as primary producers in aquatic ecosystems. Most algae are eukaryotic, photosynthetic organisms that live in a wet environment. They are distinguished from the higher plants by a lack of true roots, stems or leaves. Many species are single-celled and microscopic (including phytoplankton and other microalgae); many others are multi behavior to one degree or another, some of these growing to large size (for example, seaweeds such as kelp and *Sargassum*).

Phytoplankton

(Chlorophyceae – GA / Cyanophyceae – BGA / Bacillariophyceae – Diatoms/ flagellates)

Zooplankton

Protozoa (dinoflagellates and ciliates)

Crustacea

a. Copepoda Copepods - Cyclops

b. Cladocera Cladocerans - Daphnia

c. Ostracoda: Cypris

d. Rotifera Rotifers - *Brachionus sp*

Aquatic weeds

Insecta

Mollusca

Amphibia

Forage or bait Fishes

Micro Algae Selection Criteria

- A high nutritional value for both fish larvae and rotifers;
- Absence of toxicity for their predators;
- Suitable dimensions and digestibility, to be easily accepted by rotifers;
- A good capacity to adapt to standardized mass rearing conditions;
- A high rate of reproduction in artificial environments; reliability and affordability of their mass production.

Phytoplankton is usually grouped according to cell size. Picoplankton are the smallest and is identified as phytoplankton < 2 micrometers (µm) in diameter. Picoplankton are mostly prokaryotes such as cyanobacteria and prochlorophytes as well as several eukaryotic alga species. Nanoplankton is intermediate sized microalgae and range in size from 2-20 µm. Nanoplankton is dominated by phytoplankton with flagellas (e.g. cryptophytes and chrysophytes, and prymnesiophytes). Microplankton are the largest phytoplankton and include those algae > 20 µm in diameter. Microplankton is made up mostly of diatoms and dinoflagellates. plankters can account for 75 % of the total primary productivity of a system Tiny phytoplankters can account for as much as 50 % of the primary production in oceanic waters Phytoplankton are the food source for numerous other organisms, especially the zooplankton. Zooplankton grazers can significantly decrease phytoplankton density.

Marine Phytoplankton

In the early critical stages of the rearing larvae of finfishes and shellfishes, the phytoflagellates (species of *Isochrysis, Pavlova, Dicrateria, Chromulina and Tetraselmis*), the diatoms (species of *Chaetoceros, Skeletonema* and *Thalassiosira*) form the primary food and other nanoplankters (species of *Chlorella and Synechocystis*) form the basic food.

Isochrysis (Micro Alga) - Haptophytes

It is an outstanding food for various bivalve larvae and is now widely cultured for use in the bivalve aquaculture industry. Isochrysis is a small golden/ brown flagellate that is very commonly used in the aquaculture industry. It is high in DHA and often used to enrich zooplankton such as rotifers or Artemia. Isochrysis is primary algae used in shellfish hatcheries and used in some shrimp hatcheries. In finfish hatcheries, it is used for DHA enrichment of rotifers. It is also a good size for feeding brine shrimp and copepods. In shellfish hatcheries, Isochrysis is the most commonly used algae for oysters, clams, mussels and scallops. In shrimp hatcheries, it is used for DHA enrichment. It includes species such as Isochrysis galbana, Isochrysis litoralis and Isochrysis maritima.

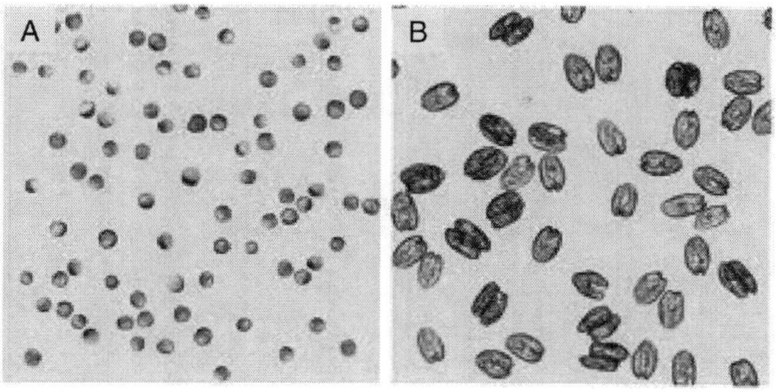

Isochrysis

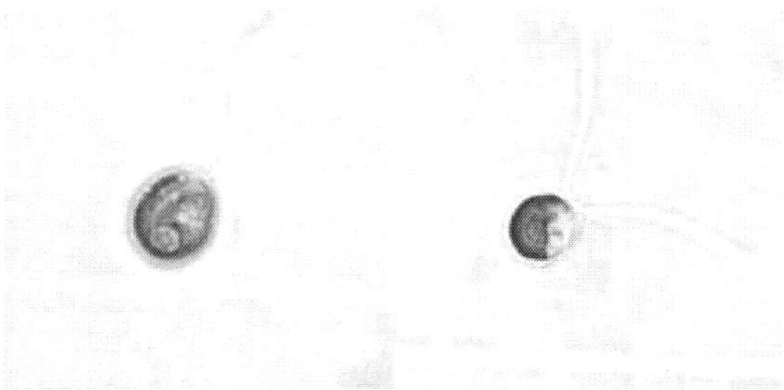

Dicrateria is a genus of haptophytes, comprising 3 species *Dicrateria gilva, Dicrateria inornata*, & *Dicrateria vlkianum*.

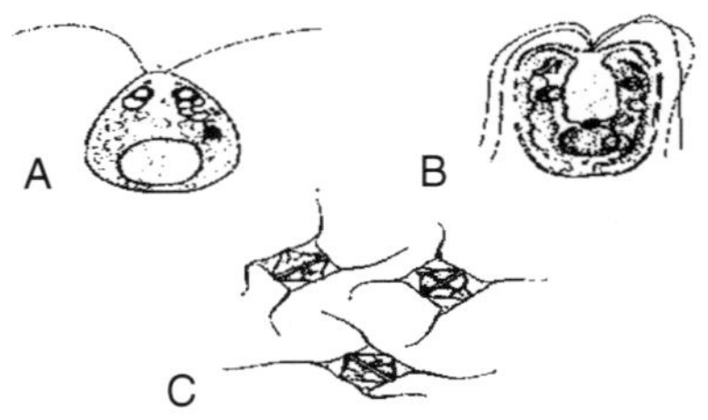

(a) *Tetraselmis* spp. (b) *Dunaliella* spp. (c) *Chaetoceros* spp.

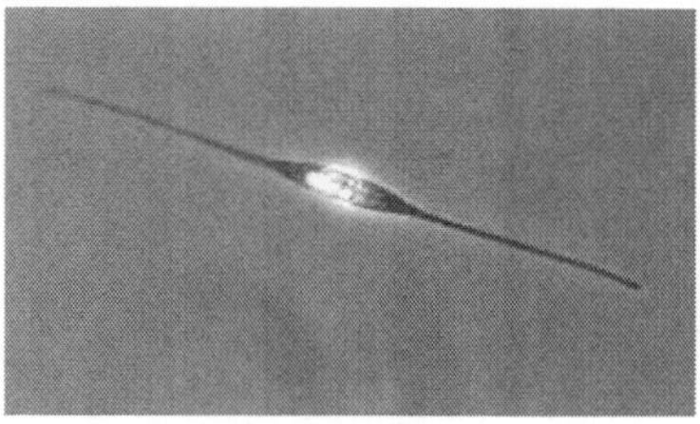

***Nitzschia* sp.**
Long, oval cell with two projections; very small cells that move in straight line

Chaetoceros

Chaetoceros is probably the largest genus of marine planktonic diatoms with approximately 400 species described. Although a large number of these descriptions are no longer valid. It is often very difficult to distinguish between different Chaetoceros species Several attempts have been made to restructure this large genus into subgenera and this work is still in progress However, most of the effort to describe species have been focused in boreal areas and the genus is cosmopolitan. So, there are probably, a large number of tropical species still undescribed.

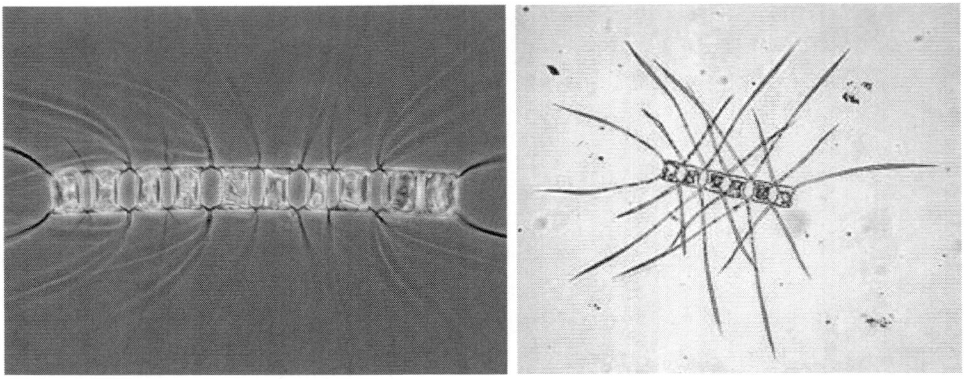

Chaetoceros

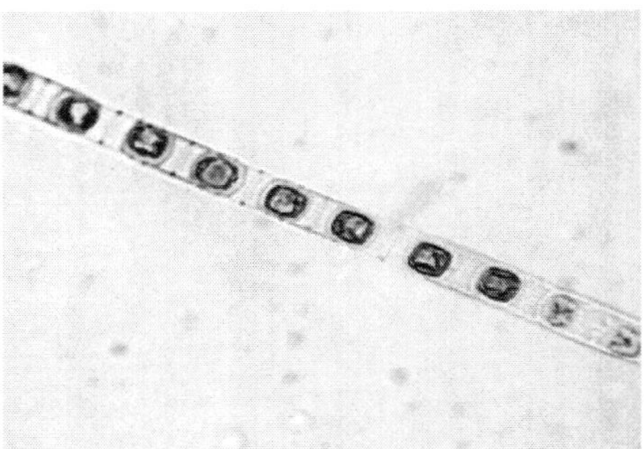

Skeletonema Costatum

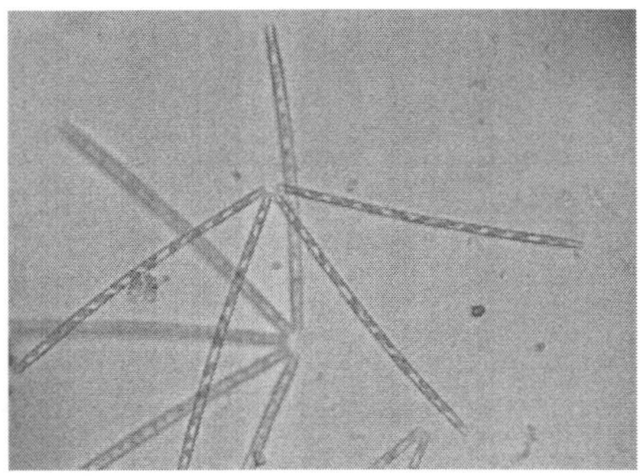

Thalassiosira weissflogii is a species of centric diatom, a unicellular microalga. It is found in marine environments and also in inland waters in many parts of the world

KINGDOM	:	*Chromalveolata*
PHYLUM	:	*Heterokontophyta*
CLASS	:	*Coscinodiscophyceae*
ORDER	:	*Thalassiosirales*
FAMILY	:	*Thalassiosiraceae*
GENUS	:	*Thalassiosira*
SPECIES	:	*T. weissflogii*

Thalassiosira weissflogii

Chlorella

Chlorella is a genus of single-cell green algae, belonging to phylum Chlorophyta. It is spherical in shape, about 2 to 10 µm in diameter and is without flagella. Chlorella contains green photosynthetic pigments chlorophyll a and b in its chloroplast. Through photosynthesis, it multiplies rapidly, requiring only carbon dioxide, water, sunlight, and small amount of minerals to reproduce Name Chlorella is taken from Greek chloros, meaning green and the Latin diminutive suffix ella, meaning small.

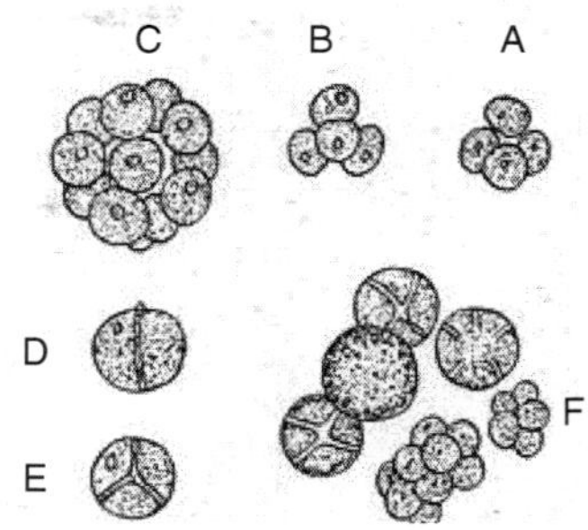

Micro algae culture- important species used as live feed

Algal Classes	*Microalgae*
Chlorophyceae	*Chlorella spp., Nannochloropsis salina, Tetraselmis spp., Dunaliella spp.*
Bacillariophyceae (Diatoms)	*Chaeloceros spp., Skeletonema costarum Thalassiosira subtilis*
Chrysophyceae (Golden yellow flagellates)	*Dicrateria inornata / Chryptochrysis sp*
Haptophyceae (Golden brown flagellates)	*Isochrysis galbana, Pavlova lutheri Chromulina frieburgensis*
Cyanophyceae (Blue green algae)	*Spirulina spp (freshwater forms) Synechocystis spp.*
Mixed culture (for feeding post larvae and spat)	*Phytoflagellates, Nanoplankters, Centric and Pennate diatoms*

Feeding protocol of micro algae in various hatcheries

Growth stage of animal	*Mieroalgae feed*
Prawn Hatchery Zoea I to Mysis I	*Chaetoceros spp, SkeLetonema costatum*
Prawn Hatchery Mysis IT to post larvae V	*Chaetoceros* + *Tetraselmis* + Rotifer + Artemia cysts
Prawn Hatchery PLVtoPLXX	Mixed culture + Rotifers + *Chaetoceros spp* + Artemia cysts
Fish Hatchery Larvae to post larvae	Phytoflagellates + Diatoms + Artemia cysts + Spirulina powder (for fresh water forms)
Molluscan Hatchery 'D' shape to spat	*Isochrysis, Pavlova, Dicrateria, Chromulina mulilla Chaetoceros spp, Skeletonema, Thaiassiosira*
Molluscan Hatchery Spat to juvenile	*Tetraselmis* + other flagellates
Molluscan Hatchery luvenile to adult	Mixed culture dominated by diatoms, *Chorella, T etraselmis* etc
Sea cucumber Hatchery Larvae to auricularia stage	Phytoflageilates + Nannoplankters

Growth stage of animal	Mieroalgae feed
Sea cucumber Hatchery Auricularia to pentacula	*Chaetoceros* + Mixed culture
Sea cucumber Hatchery Pentacula to young one	Algal powder + detritus

T-Isochrysis Algae Paste

Isochrysis is a small golden-brown flagellate commonly used in the aquaculture industry. It is high in DHA and often used to enrich zooplankton such as rotifers or artemia. Isochrysis is a primary algae found in shellfish hatcheries and in some shrimp hatcheries.This product is now glycerine free. Store in refrigerator. (Food Grade Glycerine can be added upon special request to prevent freezing by calling 801-782-4700)

Pavlova Algae Paste (125 ml)

Pavlova has a high DHA and EPA content and is excellent for enriching rotifers and other zooplankton. Pavlova also works very well for oysters, clams, mussels, and scallops.

Tetraselmis Algae Paste

Tetraselmis is a large green flagellate with a very high lipid level. It also contains natural amino acids that stimulate feeding in marine animals. It is an excellent feed for larval shrimp.

Nannochloropsis Algae Paste ("Greenwater")

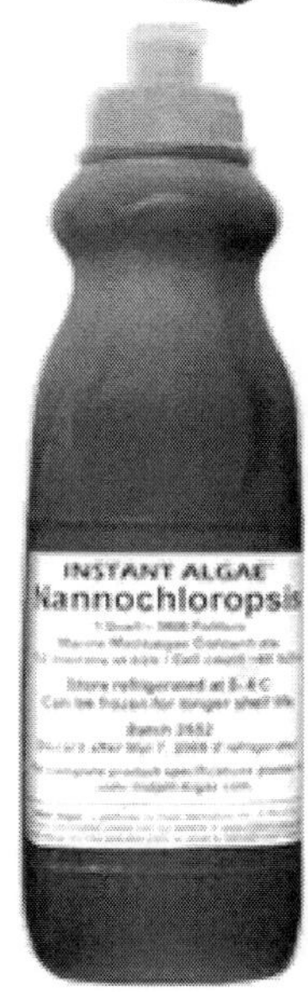

Nannochloropsis sp. is small green algae, 2 - 4 microns in diameter that is used extensively in aquaculture industry for growing small zooplankton such as rotifers, copepods, daphnia and artemia. It is also used extensively in reef tanks for feeding SPS corals and other filter feeders that require extremely small phytoplankton to thrive. General useage guideline: 1 - 2 drops for every 10 gallons aquaria.

Available in a convenient 125 ml for hobbyists NOTE: Off the shelf Nannochloropsis Algae Paste no longer contains food grade glycerine as a cryopreservative to prevent the algae from freezing solid when kept at below freezing temperatures. For daily use, Nannochloropsis Algae Paste should be stored in a refrigerator (32 to 37^{0} F) to maintain its optimum nutritional value and freshness for up to a period of 3 months. The product can also be kept frozen (as a solid) for long term storage.

Concentration: Approx. 70 billion cells / ml.

Contents: Non viable *Nannochloropsis* sp. algal cells, water.

Proximate Analysis: 35 % Protein, 7.5 % CHO, 18 % Lipid.

Freshwater Phytoplankton

Spirulina spp (freshwater forms)

Spirulina is a genus of cyanobacteria. As a photosynthetic organism capable of fixing nitrogen, its non-protein nutrient values are relatively low

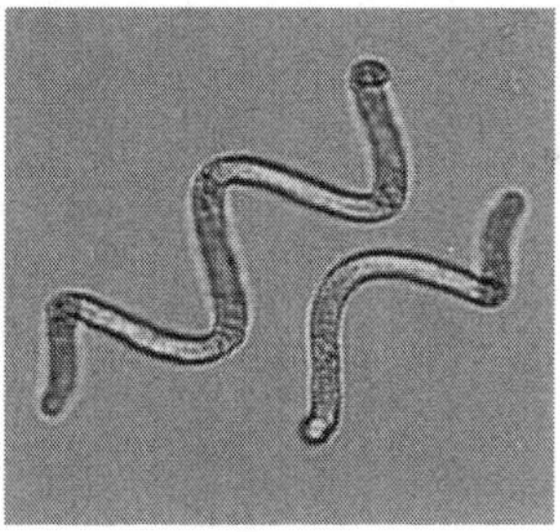

***Spirulina* (Genus)**

Marine Zooplankton

Rotifers and cladocerans are important components of most freshwater communities

Two copepod species *viz. Euterpina acutifrons* and *Pseudodiaptomus serricaudatus* employed as starter feed for the larviculture of three species of damsel fishes *viz.* the three spot damselfish, *Dascyllus trimaculatus,* the humbug damselfish, *Dascyllus aruanus* and the blue damselfish, *Pomacentrus caeruleus*.

Freshwater Zooplankton

Calanoid Copepod

Cladocera is an order of small crustaceans commonly called water fleas. Around 620 species have been recognised so far, with many more undescribed. They are ubiquitous in inland aquatic habitats, but rare in the oceans.

Daphnia is a frequently used food source in the freshwater larviculture (i.e. for different carp species) and in the ornamental fish industry (i.e. guppies, sword tails, black mollies and plattys etc.). *Daphnia* belongs to the suborder Cladocera, which are small crustaceans that are almost exclusively living in freshwater.

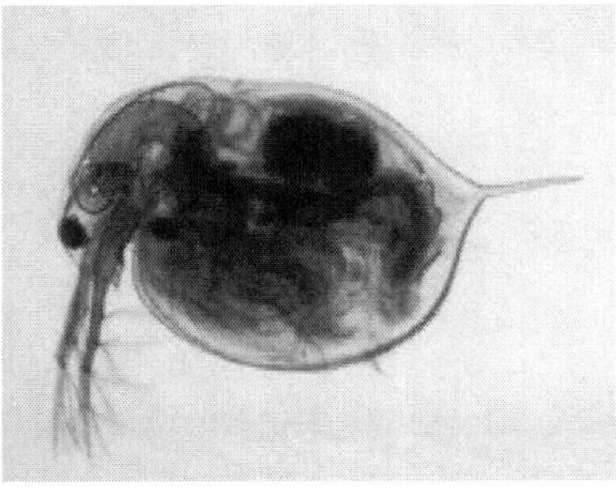

Daphnia magna

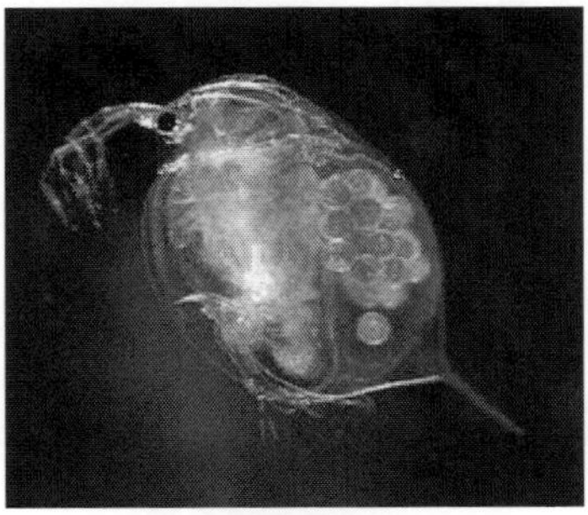

Female adult

Cladoceran (*Moina* sp.)

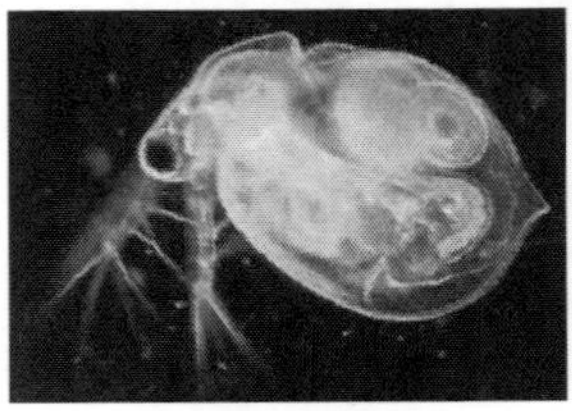

Adult Moina

Moina can be collected from brackishwater channels, pools and freshwater ponds. Moina is a genus of crustaceans within the family Moinidae. The genus was first described by W. Baird in 1850. This genus demonstrates the ability to survive in waters containing high salinity and other impurities, including salt pans.

KINGDOM	:	*Animalia*
PHYLUM	:	*Arthropoda*
SUBPHYLUM	:	*Crustacea*
CLASS	:	*Branchiopoda*
ORDER	:	*Cladoceran*
FAMILY	:	*Daphniidae*
GENUS	:	*Daphnia*
SPECIES	:	*D. magna*

Daphnia magna

Rotifer

Brachionus calyciflorus and Brachionus rubens are the most commonly cultured rotifers in freshwater mass cultures. Rotifers are small (50 - 1000 μm) zooplankton that occurs in freshwater, brackish and marine environments. Rotifers feed on microalgae and are consumed by a wide variety of fish, shellfish, corals and other organisms. They are used extensively in aquaculture and aquariums because of their very high reproductive rates (as great as doubling or better every 24 hours), ease of culturing, optimal size for larval fish, and nutritional profile that can be tailored to the needs of prey species by use of special feeds such as Reed Mariculture's RotiGrow ® and N - Rich ® microalgae feeds. Euryhalilne rotifer *B. plicatilis*; The saline rotifers - Worldwide use of the brackish water rotifers *Brachionus plicatilis* and *B. rotundiformis*; filter feeders which consume feed-particles of less than 5 microns in size. The rotifers show worldwide distribution and the majority of them inhabit freshwater and some genera also occur in brackishwater and marine habitats. The culture of *Brachionus*, in particular, *B. plicatilis* has been well documented

Brachionus plicatilis

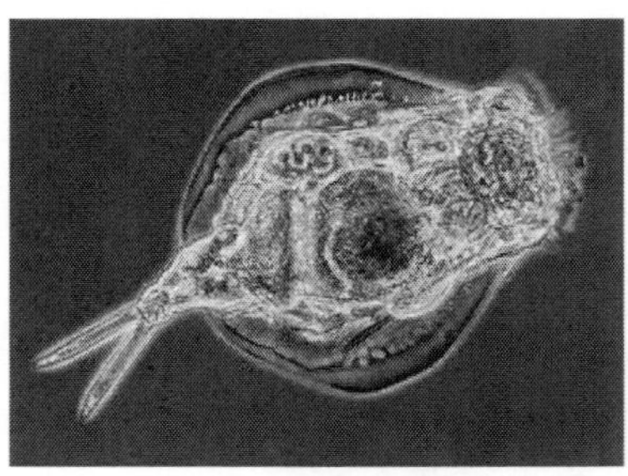

Brachionus Plicatilis **(Muller, 1786)**

Brachionus plicatilis is a euryhaline (tolerate a wide range of salinity) rotifer in the Family Brachionidae and is possibly the only commercially important rotifer, being raised in the aquaculture industry as food for fish larvae. It has a broad distribution in salt lakes around the world and has become a model system for studies in ecology and evolution. DNA sequencing studies in 2002 revealed *Brachionus plicatilis* to be a cryptic species complex comprising approximately 10 separate species, each of which has been diverging for several million years. The species, rather than the complex, is now usually referred to as Brachionus plicatilis sensu stricto. Haploid '1C' genome sizes in the Brachionus plicatilis species complex range at least from 0.056 to 0.416 pg.

KINGDOM : *Animalia*
PHYLUM : *Rotifera*
CLASS : *Monogononta*
ORDER : *Ploimida*
FAMILY : *Brachionidae*
GENUS : *Brachionus*
SPECIES : *B. plicatilis*

Freshwater Zooplankton

Cyclops

The copepod Cyclops serulatus and the cladoceran Simocephalus serrulatus were the most widely distributed taxa. Cyclops is one of the most common genera of freshwater copepods, comprising over 400 species. Cyclops has a cosmopolitan distribution in fresh water, but is less frequent in brackish water.

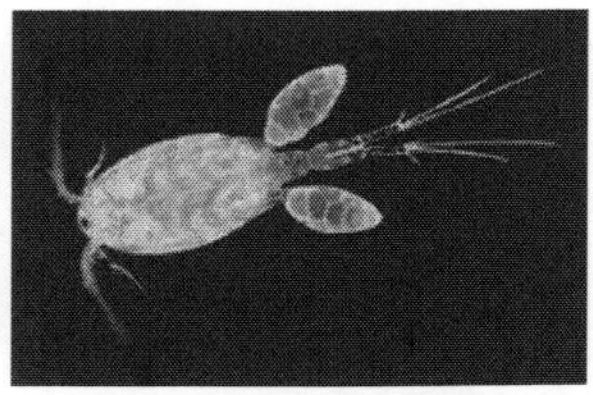

Cyclops bicuspidatus

Major classes and genera of micro algae cultured in aquaculture (Modified from De Pauw and Persoone, 1988)

Class	*Genus*	*Examples of application*
Bacillariophyceae	*Skeletonema*	PL, BL, BP
	Thalassiosira	PL, BL, BP
	Phaeodactylum	PL, BL, BP, ML, BS
	Chaetoceros	PL, BL, BP, BS
	Cylindrotheca	PL
	Bellerochea	BP
	Actinocyclus	BP
	Nitzchia	BS
	Cyclotella	BS
Haptophyceae	*Isochrysis*	PL, BL, BP, ML, BS
	Pseudoisochrysis	BL, BP, ML
	Dicrateria	BP
Chrysophyceae	*Monochrysis (Pavlova)*	BL, BP, BS, MR
Prasinophyceae	*Tetraselmis (Platymonas)*	PL, BL, BP, AL, BS, MR
	Pyramimonas	BL, BP
	Micromonas	BP

Cryptophyceae	*Chroomonas*	BP
	Cryptomonas	BP
	Rhodomonas	BL, BP
Cryptophyceae	*Chlamydomonas chlorococcum*	BL, BP, FZ, MR, BS BP
Xanthophyceae	*Olisthodiscus*	BP
Chlorophyceae	*Carteria*	BP
	Dunaliella	BP, BS, MR
Cyanophyceae	*Spirulina*	PL, BP, BS, MR

PL, penaeid shrimp larvae

BL, bivalve mollusc larvae

ML, freshwater prawn larvae

BP, bivalve mollusc postlarvae

L, abalone larvae

MR, marine rotifers (*Brachionus*)

BS, brine shrimp (Artemia)

SC, saltwater copepods

FZ, freshwater zooplankton

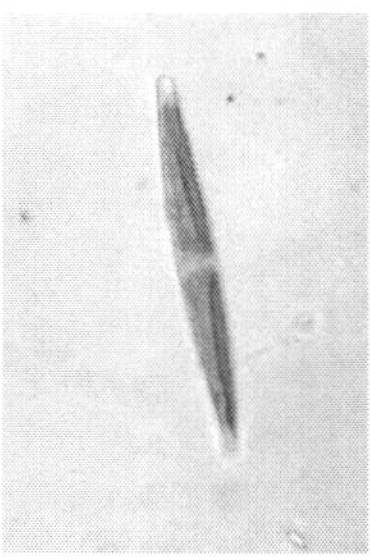

Nitzchia

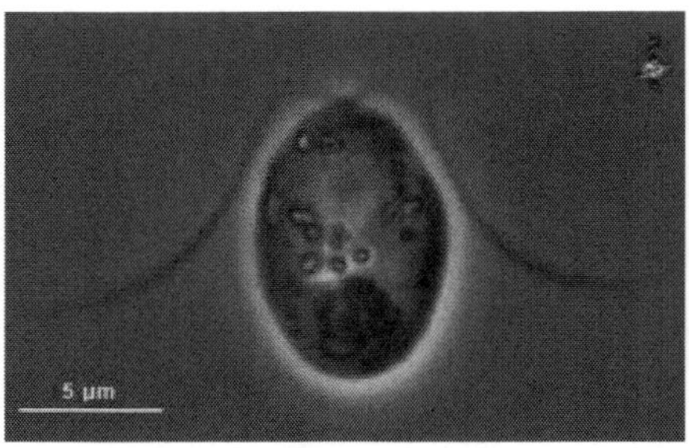

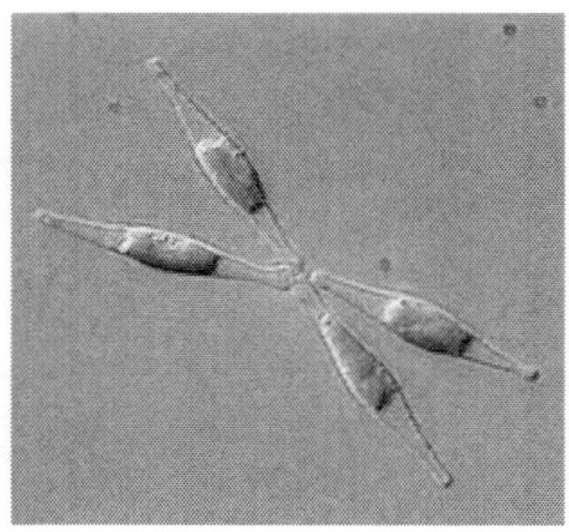

Phaeodactylum

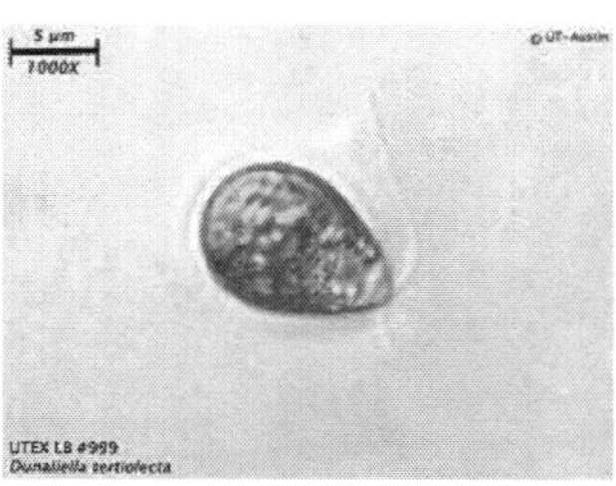

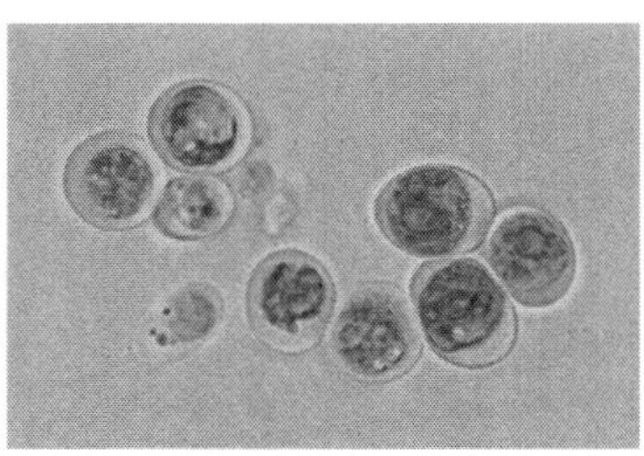

Chlamydomonas chlorococcum

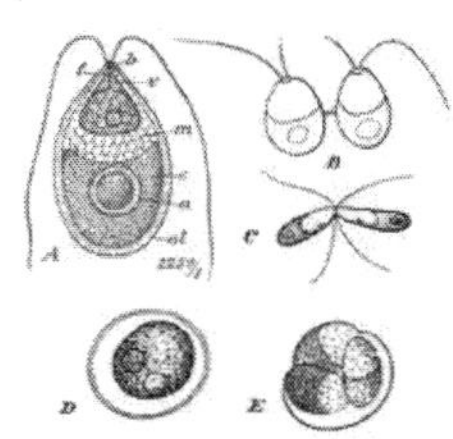

Dunaliella

Chapter 6

Current Status of Live Feed Production for Larviculture

At World Level

Since 1960's, large scale production of marine finfish is pioneered in Japanese red sea bream (*Pagrus major*).

In 1960's, Europe and USA undertook research in Marine Fish Larviculture.

Early 1980's, British and French scientist, setting up of first commercial hatchery For European seabass (*Dicentrachus labrax*), Gilthead seabream (*Sparus aurata*) and the turbot (*Scophthalamus maximus*).

Today, Japan is still the biggest producer of marine fish fry of 200 million fry/ year. Red sea bream (*Pagrus major*) and Japanesh flounder (*Paralichthys olivaceus*) makes up 70 % of total production.

Europe comes next. In 1993, Hatcheries in the Mediterranean's produced more than 100 million fry / year (Bass and Bream).

Very significant was achieved in the last decades in improving nutrition particularly through manipulation of fatty acid profile of live prey Brachionus and the Artemia. However, the most widely used item is Artemia. Of which, today more than 2000 rats of cysts are marketed annually for feeding fin and shell fish hatcheries. It is more clear that in larval nutrition. Most attention given was qualitative and qualitative HUFA requirements. Further study is required on ratios of n3 / n6 / n9 fatty acids particularly the effects of the lipids except phospho lipids (phospho lipid affects which might alter the requirement for n3 HUFAs). Apart from the above, many other dietary compounds deserve attention such as vitamins, free amino acids, pigments etc. The magic effect of green water also requires further study. In the intensive culture of marine fish larvae, a microalga is often added to the water along with microzooplankton or microparticulate food (the "green water" technique). The application of cofeeding principle that is supplementing live feed with formulated feeds

Although still room for improvements particularly cost effectiveness, present day hatchery technology for marine fish species has proven to be widely applicable both geographical location and fish species The list of commercially cultivated species is growing fast such as Bass, Seabream, Turbot, Flounder, Fugu, Milk fish, Mullet and many candidate species will hopefully will join the above list such as dolphin fish (mahi-mahi), various grouper species, red snapper, halibut, wolfish.

India

The major expansion of marine finfish and crustacean aquaculture since 1980s around the world can be attributed to the development of standard mass production techniques of live feeds. Even though, most farmed marine animals are either carnivorous or omnivorous from their post larval stages, micro algae are required for larval nutrition during a brief period either for direct consumption or indirectly as food for live prey fed to small larvae. The hatchery production of penaeid shrimp post larvae depends on the use of live diatoms for the early stages and Artemia for later stages. The hatchery production of juveniles of marine finfish is achieved globally by the use of rotifers and Artemia. Microalgae are also routinely used in the 'greenwater technique' employed for marine finfish larviculture. Most marine finfishes have altricial larvae and when yolk sac is exhausted, they remain in an undeveloped state. The digestive system is rudimentary, lacking a stomach and much of the protein digestion takes place in the hindgut epithelial cells. Altricial larvae cannot digest formulated feeds and hence live feed is vital for their survival. Live feeds are able to swim in water column and are thus constantly available to the larvae. The movement of live feed in water stimulates larval feeding responses.

Micro Algae

Micro algae play a critical role in the marine aquaculture of molluscs, shrimps and fish. Nearly 16 genera of micro algae are commonly employed for aquaculture purposes. They are generally free living, pelagic and in the nanno plankton range (2 – 20 µm). Mass culture, batch cultures are used for propagation of micro algae. Even after the endo- exotrophic phase, micro algae have a positive effect on larviculture and may increase the resistance of larvae to further stressing or adaptive conditions. The indirect effects of micro algae on larvae are mainly related to water quality, luminosity, the bacteriology of water the quality and accessibility of rotifers.

It is thus evident that strategic use of micro algae in hatcheries during the very early life of marine fish improves the success of first feeding, a prerequisite for efficient survival, growth and quality in fish larviculture.

Rotifers

Since four decades, Rotifers have been used as live feed for cultured marine fish. It is well known that a continuous, stable and reliable supply of nutritionally adequate rotifers is the key to the larviculture of marine finfish. Rotifers of the species such as *Brachionus rotundiformis, B. plicatilis* almost indispensable for larval rearing of most marine finfish. The success of rotifer cultivation is dependent on selecting the most suitable rotifer species or strain for local culture conditions maintaining water quality in culture tanks choosing the most appropriate culture technique.

Since larval production depends on a predictable and reliable supply of rotifers, evaluating the physiological state of rotifer culture is very important in hatcheries.

Six parameters viz egg ratio, swimming velocity, ingestion rate, viscosity, enzyme activity and diseases employed for assessing the state of health of rotifer cultures. The nutritional quality of rotifers is improved by enrichment Rotifers are collected or harvested from culture tanks into containers. They are kept at very high densities and incubated for 8 - 20 hours with enrichment dietary components like HUFA. In addition to nutritional enrichment, Rotifers can be enriched with antibiotics or with probiotic bacteria. It is evident that rotifer cultures will continue to be indispensable in marine finfish hatcheries. Current methodologies of producing and enriching rotifers are meeting the requirements of the industry. The current need to have very small sized rotifers is difficult to achieve, although several super small strains have been found and cultured. Improved methods for predicting the health of cultured rotifers may be useful in preventing culture crashes. Using preserved rotifers may eliminate the dependence on daily production of rotifers. Cheaper methods of resting egg production are another field which requires research attention in future.

Copepods

Copepods are a major component of the natural diet of marine fish larvae. The advantages of copepods over rotifers are that copepods have wide range of body sizes both within and between species. The early stage nauplii and copepodites can be extremely useful as initial prey for species that have very small larvae with small mouth gape at first feeding. In extensive methods copepods are collected from nature and inoculated into outdoor tanks to mass produce for fish larval rearing. Improved growth, survival and / or rates of normal pigmentation have been documented for several marine fish species fed copepods alone or as a supplement to other traditional live feeds. The improvements in larval growth, survival and normal pigmentation are generally attributed to the levels of DHA, EPA and / or arachidonic acid (ARA) in the diet and in particular to the DHA : EPA ratio in the diet. Copepods which constitute the major diet for marine fish larvae in nature contain high levels of DHA and other PUFA. DHA levels in wild copepods can be more than 10 times higher than in enriched Artemia

Artemia

Nauplii in instar I and II stages are the most widely used forms of Artemia in aquaculture. They are the earliest and easiest live feed, being obtained directly from the cysts. Artemia Widely used in the mass culture of different sea bream species, sea bass species, wolf fish, cod, turbot, halibut, flounder species, milk fish, surgeon and many shrimps, prawn, crabs and lobsters. Several factors are critical for the successful hatching of Artemia cysts. Optimal hatching conditions are constant temperature, 15 - 35 g / l salinity, pH around 8.0, Minimum oxygen levels of 2 mg / l, preferably 5 mg / l, maximum cyst densities of 2 g / l, and strong illumination (2000 lux). Best hatching results are achieved in containers with conical bottom, aerated from the bottom. Transparent or translucent containers will facilitate inspection of hatching, especially when harvesting. After hatching and before feeding to fish larvae, the nauplii should be separated from the hatching wastes (empty cyst shells, unhatched cyst, debris, microorganism and hatching metabolites). Decapsulation of cysts results in disinfection of the cysts and also eliminates the introduction of cyst shells to culture tanks. Most marine fish larvae cannot synthesize DHA, EPA or Arachidonic acid from shorter chain precursors. They must be provided in the larval dietHence artemia is enriched for enhancing the nutritional value for using as a live feed. Although artemia is often an inferior food source for fish larvae compared with wild zooplankton, the ability to produce any amount of biomass within 24 hrs, the constant improvement of enrichment products ensure its continued use in marine fish larviculture. It is quite possible that artemia will gradually replaced by formulated diets It is obvious that the use of nauplii will continue in hatcheries at least for a few more years.

Prospects

It is quite obvious that at the present status of technology, live feeds cannot be dispensed in marine larviculture. The major challenge for the future is in the development of commercial artificial feeds for larvae. The immediate research thrust should be on technologies for high density and stable production of high quality live feeds. The identification of causes for sudden crashes in culture and remedial measures for the same requires priority attention. Genetic manipulations for the production of super small strains of rotifers and the commercial production of rotifer cysts are also major areas for intensified research. The production of high density microalgae in photo bioreactors and by fermentation techniques are potential areas for future development. The copepod culture techniques, which is the key to the success of rearing of altricial larvae of marine finfish still remains to be standardized. Hence this should receive more focused research than the better studied traditional live feeds.

Future Scope of Micro Algal Culture

The development of mass culture of micro algae offers immense scope and can run concurrently with mariculture programmes. At the same time it offers a challenge for the phycologists. Considerable research is needed for elucidation of the key problem such as the growth kinetics of different species, the period of economically viable harvest, etc. In the matter of waste water recycling, unicellular algae hold out much promise. Maximization of solar energy utilization under controlled conditions of temperature, light and aeration, development of viable methods for intensive culture, maintenance, formation of extracellular products and their significance are promising lines of investigations. Most of the farm sites and marketing centres are far apart. Majority of the fishermen are below the poverty line. Therefore, a well planned strategy to synchronize the various aspects of development such as water supply, electricity, approachable road, and marketing system has to be envisaged. Training of technical manpower, inputs and basic technology also should be looked into. Conventional aquaculture practices have been in existence. New technologies such as the growing of food organisms for the fishmeal production centres should be introduced. What is urgently required is to raise this technology at the level of organized small-scale industries and spread them to all parts of the country.

Chapter 7

Nutritional Properties of Microalgae

Micro algal species can vary significantly in their nutritional value, and this may also change under different culture conditions. Nevertheless, a carefully selected mixture of microalgae can offer an excellent nutritional package for larval animals, either directly or indirectly (through enrichment of zooplankton). Microalgae that have been found to have good nutritional properties – either as monospecies or within a mixed diet - include *C. calcitrans*, *C. muelleri*, *P. lutheri*, *Isochrysis* sp. (T.ISO), *T. suecica*, *S. costatum* and *Thalassiosira pseudonana*. Several factors can contribute to the nutritional value of a microalga, including its size and shape, digestibility (related to cell wall structure and composition), biochemical composition (eg. nutrients, enzymes, toxins if present) and the requirements of the animal feeding on the alga. Since the early reports that demonstrated biochemical differences in gross composition between microalgae and fatty acids, many studies have attempted to correlate the nutritional value of microalgae with their biochemical profile. However, results from feeding experiments that have tested microalgae differing in a specific nutrient are often difficult to interpret because of the confounding effects of other micro algal nutrients. Nevertheless, from examining all the literature data, including experiments where algal diets have been supplemented with compounded diets or emulsions, some general conclusions can be reached. Microalgae grown to late-logarithmic growth phase typically contain 30 to 40 % protein, 10 to 20 % lipid and 5 to 15 % carbohydrate. When cultured through to stationary phase, the proximate composition of microalgae can change significantly; for example when nitrate is limiting, carbohydrate levels can double at the expense of protein. There does not appear to be a strong correlation between the proximate composition of microalgae and nutritional value, though algal diets with high levels of carbohydrate are reported to produce the best growth for juvenile oysters (*Ostrea edulis*) and larval scallops (*Patinopecten yessoensis*) provided polyunsaturated fatty acids (PUFAs) are also present in adequate proportions. In contrast, high dietary protein provided best growth for juvenile mussels (*Mytilus trossulus*) and Pacific oysters (*Crassostrea gigas*).

PUFAs derived from microalgae, i.e. docosahexaenoic acid (DHA), eicosapentaenoic acid (EPA) and arachidonic acid (AA) are known to be essential for various larvae. The fatty acid content showed systematic differences according to taxonomic group, although there were examples of significant differences between microalgae from the same class. Most micro algal species have moderate to high percentages of EPA (7 to 34%; Fig 1). Prymnesiophytes (eg. *Pavlova* spp. and *Isochrysis* sp. (T.ISO)) and cryptomonads are relatively rich in DHA (0.2 to 11 %), where as eustigmatophytes (*Nannochloropsis* spp.) and diatoms have the highest percentages of AA (0 to 4 %). Chlorophytes (*Dunaliella* spp. and *Chlorella* spp.) are deficient in both C20 and C22 PUFAs, although some species have small amounts of EPA (up to 3.2 %). Because of this PUFA deficiency, chlorophytes generally have low nutritional value and are not suitable as a single species diet. Prasinophyte species contain significant proportions of C20 (*Tetraselmis* spp.) or C22 (*Micromonas* spp.) - but rarely both. While the importance of PUFAs is recognized, the quantitative requirements of larval or juvenile animals feeding directly on microalgae are not well established. The growth of Pacific oyster *C. gigas* larvae was not improved by feeding them microalgae containing higher than 2 % (total fatty acids) of DHA; moreover the percentage of dietary EPA was negatively correlated to larval growth. However, the authors found a correlation between the percentage composition of the short chain fatty acids 14:0 + 16:0 in microalgae, and larval growth rates. They reasoned that diets with higher percentages of the saturated fats were more beneficial for the rapidly growing larvae, because energy is released more efficiently from saturated fats than unsaturated fats. In late-logarithmic phase, prymnesiophytes, on average, contain the highest percentages of saturated fats (33% of total fatty acids), followed by diatoms and eustigmatophytes (27%), prasinophytes and chlorophyte s (23 %) and cryptomonads (18%) (Brown *et al.*, 1997). The content of saturated fats in microalgae can also be improved by culturing under high light conditions. The content of vitamins can vary between microalgae. Ascorbic acid shows the greatest variation, i.e. 16 fold (1 to 16 mg g / l dry weight (Brown & Miller, 1992). Concentrations of other vitamins typically show a two to four-fold difference between species, i.e. □-carotene 0.5 to 1.1 mg / g, niacin 0.11 to 0.47 mg / g, □-tocopherol 0.07 to 0.29 mg / g, thiamin 29 to 109 µg / g, riboflavin 25 to 50 µg / g, pantothenic acid 14 to 38 µg / g, folates 17 to 24 µg / g, pyridoxine 3.6 to 17 µg / g, cobalamin 1.8 to 7.4 µg / g, biotin 1.1 to 1.9 µg / g; retinol ≤ 2.2 µg /g and vitamin D < 0.45 µg /. To put the vitamin content of the microalgae into context, data should be compared with the nutritional requirements of the consuming animal. Unfortunately, nutritional requirements of larval or juvenile animals that feed directly on microalgae are, at best, poorly understood. However, the requirements of the adult are far better known (eg. For marine fish and prawns) and in the absence of information to the contrary, will have to serve as a guide for the larval animal. These data suggest that a carefully selected, mixed algal diet should provide adequate concentrations of the vitamins for aquaculture food chains. The amino acid composition of the protein of microalgae is very similar between species and relatively unaffected by the growth phase and light conditions. Further, the composition of essential amino acids in microalgae is very similar to that of protein from oyster larvae (*C. gigas*). This indicates that it is unlikely the protein quality is a factor contributing to the differences in nutritional value of microalgal species. Sterols, minerals and pigments also may contribute to nutritional differences of microalgae.

Nutritional Value of Micro-Algae

The nutritional value of any algal species for a particular organism depends on its cell size, digestibility, production of toxic compounds, and biochemical composition. The gross composition of 16 species of micro-algae is compared in Table. Although there are marked differences in the compositions of the micro-algal classes and species, protein is always the major organic constituent, followed usually by lipid and then by carbohydrate. Expressed as percentage of dry weight, the range for the level of protein, lipid and carbohydrate are 12 - 35 %, 7.2 – 23 % and 4.6 – 23 % respectively. The content of highly unsaturated fatty acids (HUFA), in particular eicosapentaenoic acid (20:5 n-3, EPA), arachidonic acid (20:4 n-6, ARA), and docosahexaenoic acid (22:6 n-3, DHA), is of major importance in the evaluation of the nutritional composition of an algal species to be used as food for marine organisms. The fatty acid composition of 10 species of micro-algae grown under defined conditions and harvested during the log phase. Significant concentrations of EPA are present in the diatom species (*Chaetoceros calcitrans, C. gracilis, S. costatum, T. pseudonana*) and the prymnesiophyte *Platymonas lutheri*, whereas high concentrations of DHA are found in the prymnesiophytes (*P. lutheri, Isochrysis* sp.) and *Chroomonas salina*. Micro algae can also be considered as a rich source of ascorbic acid (0.11 - 1.62 % of dry weight. The nutritional value of micro algae can vary considerably according to the culture conditions. For example, the effect of the composition of the culture medium on the proximate composition of various species of micro-algae is demonstrated in the following Table. The protein content per cell, which is considered as one of the most important factors determining the nutritional value of micro-algae as feed in aquaculture, was found to be more susceptible to medium-induced variation than the other cellular constituents.

Concentrations of chlorophyl a, protein, carbohydrate and lipid in 16 species of micro algae commonly used in aquaculture (Modified from Brown, 1991).

Algal class Species	*Dry weight (pg.cell-1)*	*Chl a*	*Protein*	*CHO*	*Lipid*
	Weight of constituent (pg.cell-1)				
Bacillariophyceae					
Chaetoceros calcitrans	11.3	0.34	3.8	0.68	1.8
Chaetoceros gracilis	74.8	0.78	9.0	2.0	5.2
Nitzchia closterium	-	-	-	-	-
Phaeodactylum tricornutum	76.7	0.41	23.0	6.4	10.7
Skeletonema costatum	52.2	0.63	13.1	2.4	5.0
Thalassiosira pseudonana	28.4	0.27	9.7	2.5	5.5
Chlorophyceae					
Dunaliella tertiolecta	99.9	1.73	20.0	12.2	15.0

Algal class Species	Dry weight (pg.cell-1)	Chl a	Protein	CHO	Lipid
	Weight of constituent (pg.cell-1)				
Nannochloris atomus	21.4	0.080	6.4	5.0	4.5
Cryptophyceae					
Chroomonas salina	122.5	0.98	35.5	11.0	14.5
Eustigmatophyceae					
Nannochloropsis oculata	6.1	0.054	2.1	0.48	1.1
Prasinophyceae					
Tetraselmis chui	269.0	3.83	83.4	32.5	45.7
Tetraselmis suecica	168.2	1.63	52.1	20.2	16.8
Prymnesiophyceae					
Isochrysis galbana	30.5	0.30	8.8	3.9	7.0
Isochrysis aff. Galbana (T-iso)	29.7	0.29	6.8	1.8	5.9
Pavlova lutheri	102.3	0.86	29.7	9.1	12.3
Pavlova salina	93.1	0.34	24.2	6.9	11.2
Percentage of Dry Weight					
Bacillariophyceae					
Chaetoceros calcitrans	11.3	3.01	34	6.0	16
Chaetoceros gracilis	74.8	1.04	12	4.7	7.2
Nitzchia closterium	-	-	26	9.8	13
Phaeodactylumtricornutum	76.7	0.53	30	8.4	14
Skeletonema costatum	52.2	1.21	25	4.6	10
Thalassiosira pseudonana	28.4	0.95	34	8.8	19
Chlorophyceae					
Dunaliella tertiolecta	99.9	1.73	20	12.2	15
Nannochloris atomus	21.4	0.37	30	23.0	21
Cryptophyceae					
Chroomonas salina	122.5	0.80	29	9.1	12
Eustigmatophyceae					
Nannochloropsis oculata	6.1	0.89	35	7.8	18
Prasinophyceae					
Tetraselmis chui	269.0	1.42	31	12.1	17
Tetraselmis suecica	168.2	0.97	31	12.0	10
Prymnesiophyceae					
Isochrysis galbana	30.5	0.98	29	12.9	23

Algal class Species	Dry weight (pg.cell-1)	Chl a	Protein	CHO	Lipid
	Weight of constituent (pg.cell-1)				
Isochrysis aff. Galbana (T-iso)	29.7	0.98	23	6.0	20
Pavlova lutheri	102.3	0.84	29	9.0	12
Pavlova salina	93.1	0.98	26	7.4	12

Moreover, growth of animals fed a mixture of several algal species is often superior to that obtained when feeding only one algal species. A particular alga may lack a nutrient, while another alga may contain that nutrient and lack a different one. In this way, a mixture of both algal species supplies the animals with an adequate amount of both nutrients. An extensive review of nutritional aspects of micro-algae used in mariculture of bivalve molluscs, crustaceans & fish is presented in Brown *et al.* (1989).

Cellular density (106 cells / ml) and proximate composition (pg / cell) of four marine micro-algae grown in different culture media (Algal-1 is a commercial nutrient) (Modified from Herrero et al., 1991)

	Cellular density	Protein	Carbohydrates	Lipids
T. suecica				
Walne	2.29	13.31	6.20	7.04
ES	2.58	16.98	6.93	7.22
F/2	2.38	21.75	8.37	7.92
Algal-1	4.11	32.22	8.83	8.65
D. tertiolecta				
Walne	4.04	13.37	13.22	22.28
ES	4.24	14.88	15.73	23.94
F/2	4.97	13.26	17.91	23.67
Algal-1	8.45	18.82	11.08	18.18
I. galbana				
Walne	10.11	5.17	4.28	25.95
ES	12.09	7.23	5.21	28.38
F/2	10.81	8.13	5.59	26.82
Algal-1	16.15	9.57	4.28	20.68
P. tricornutum				
Walne	19.01	2.65	6.42	6.51
ES	16.23	5.21	9.20	6.45
F/2	24.65	3.34	6.90	5.52
Algal-1	39.04	4.20	5.98	5.79

Chapter 8

Types of Algal Cultures

The various types of algal cultures (phytoplankton) include:

Maintenance Cultures - natural collections of algae kept in culture vessels in the laboratory; here, succession of the previously less abundant species over the dominant ones may occur.

Enrichment Cultures - refer to crude collections of algae or other algal source materials treated with specially selected culture media which will favour the rapid increase in number of desired algal species.

Unialgal Cultures - refer to populations consisting of a single algal species, although other micro-organisms may be associated.

Axenic Culture - contains a population of a single algal species, all other living organisms being absent.

Clonal Cultures - refer to algal populations descended asexually from a single individual. Such type may or may not be axenic.

Chapter 9

Algal Culture Techniques

Algae can be produced using a wide variety of methods, ranging from closely-controlled laboratory methods to less predictable methods in outdoor tanks. The terminology used to describe the type of algal culture includes:

Indoor/ Outdoor

Indoor culture allows control over illumination, temperature, nutrient level, contamination with predators and competing algae, whereas outdoor algal systems make it very difficult to grow specific algal cultures for extended periods.

Open/ Closed

Open cultures such as uncovered ponds & tanks (indoors / outdoors) are more readily contaminated than closed culture vessels viz tubes, flasks, carboys, bags, etc.

Axenic (Sterile)/ Xenic

Axenic cultures are free of any foreign organisms such as bacteria and require a strict sterilization of all glassware, culture media and vessels to avoid contamination. The latter makes it impractical for commercial operations. Axenic culture – growing algae with no bacteria (or viruses); take unialgal culture and either screen filter gently (allowing bacteria to pass), centrifuge gently (collecting heavy fraction), or use repeated dilutions into sterile medium and antibiotics to kill bacteria

Advantages

(1) Have pure algae

Disadvantages

(1) Very difficult and time consuming

(2) Need to check repeatedly that you don't have bacterial contaminants

(3) Not all algae have been successfully grown axenically

Major advantages and disadvantages of the various algal culture techniques

Culture type	*Advantages*	*Disadvantages*
Indoors	A high degree of control (predictable)	Expensive
Outdoors	Cheaper	Little control (less predictable)
Closed	Contamination less likely	Expensive
Open	Cheaper	Contamination more likely
Axenic	Predictable, less prone to crashes	Expensive, difficult
Non-axenic	Cheaper, less difficult	More prone to crashes
Continuous	Efficient, provides a consistent supply of high-quality cells, automation, highest rate of production over extended periods	Difficult, usually only possible to culture small quantities, complex, equipment expenses may be high
Semi-continuous	Easier, somewhat efficient	Sporadic quality, less reliable
Batch	Easiest, most reliable	Least efficient, quality may be inconsistent

Batch/ Continuous/ Semi-Continuous

These are the three basic types of phytoplankton culture which will be described in the following sections:

Batch Culture

The batch culture consists of a single inoculation of cells into a container of fertilized seawater followed by a growing period of several days and finally harvesting when the algal population reaches its maximum or near-maximum density. In practice, algae are transferred to larger culture volumes prior to reaching the stationary phase and the larger culture volumes are then brought to a maximum density and harvested. The following consecutive stages might be utilized: test tubes, 2 l flasks, 5 and 20 l carboys, 160 l cylinders, 500 l indoor tanks, 5,000 l to 25,000 l outdoor tanks

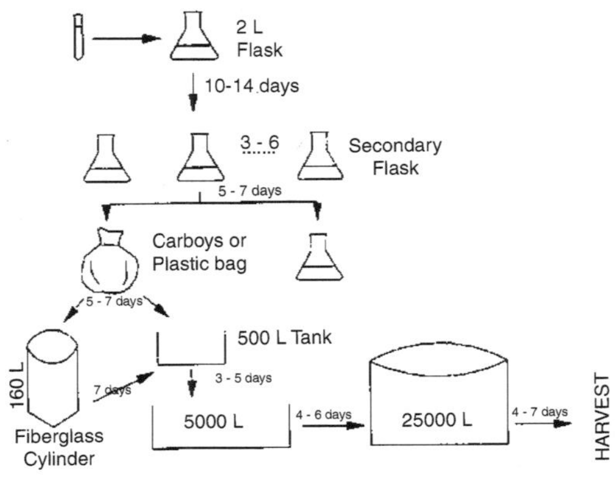

Production scheme for batch culture of algae

Batch culture systems for the mass production of micro-algae in 20,000 l tanks

Batch culture systems for mass production of micro-algae in 150 l cylinders.

According to the algal concentration, the volume of the inoculum which generally corresponds with the volume of the preceding stage in the upscaling process amounts to 2 - 10 % of the final culture volume. An inoculation schedule for the continuous production according to the batch technique is presented in Table

Inoculation schedule for the continuous production of micro-algae using the batch technique. Every week a serial is initiated with 4 or 7 test tubes, depending on whether a new culture is required for harvesting every 2 days or daily.

Days	*New culture available for harvest every 2 days*				*Harvest required daily*						
1	t	t	t	t	t	t	t	t	t	t	t
2	t	t	t	t	t	t	t	t	t	t	t
3	t	t	t	t	t	t	t	t	t	t	t
4	t	t	t	t	t	t	t	t	t	t	t
5	t	t	t	t	t	t	t	t	t	t	t
6	t	t	t	t	t	t	t	t	t	t	t
7	t	t	t	t	t	t	t	t	t	t	t
8	e	e	e	e	e	e	e	e	e	e	e
9	e	e	e	e	e	e	e	e	e	e	e
10	e	e	e	e	e	e	e	e	e	e	e
11	e	e	e	e	e	e	e	e	e	e	e
12	E	e	e	e	E	e	e	e	e	e	e
13	E	e	e	e	E	E	e	e	e	e	e
14	E	E	e	e	E	E	E	e	e	e	e
15	E	E	e	e	E	E	E	E	e	e	e

16	f	E	E	e	f	E	E	E	E	e	e
17	f	E	E	e	f	f	E	E	E	E	e
18	f	f	E	E	f	f	f	E	E	E	E
19	f	f	E	E	f	f	f	f	E	E	E
20	F	f	f	E	F	f	f	f	f	E	E
21	F	f	f	E	F	F	f	f	f	f	E
22	F	F	f	f	F	F	F	f	f	f	f
23	F	F	f	f	F	F	F	F	f	f	f
24	L	F	F	f	L	F	F	F	F	f	f
25	L	F	F	f	L	L	F	F	F	F	f
26	*	L	F	F	*	L	L	F	F	F	F
27		L	F	F		*	L	L	F	F	F
28		*	L	F			*	L	L	F	F
29			L	F				*	L	L	F
30			*	L					*	L	L
31				L						*	L
32				*							*

t = 20 ml test tube / e = 250 ml Erlenmeyer flask

E = 2 l Erlenmeyer flask / f = 30 l fiberglass tank

F = 300 l fiberglass tank

L = use for larval feeding or to inoculate large volume (> 1.5 t) outdoor tanks

* = termination of 300 l fiberglass tank

Where small amounts of algae are required, one of the simplest types of indoor culture employs 10 to 20 l glass or plastic carboys (Figure given below), which may be kept on shelves backlit with fluorescent tubes

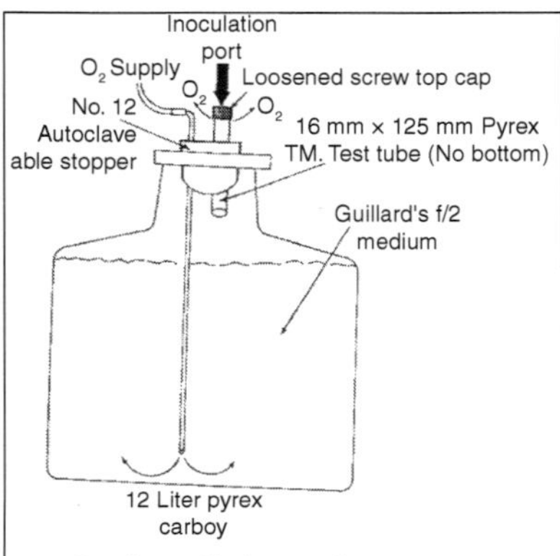

Carboy Culture Apparatus

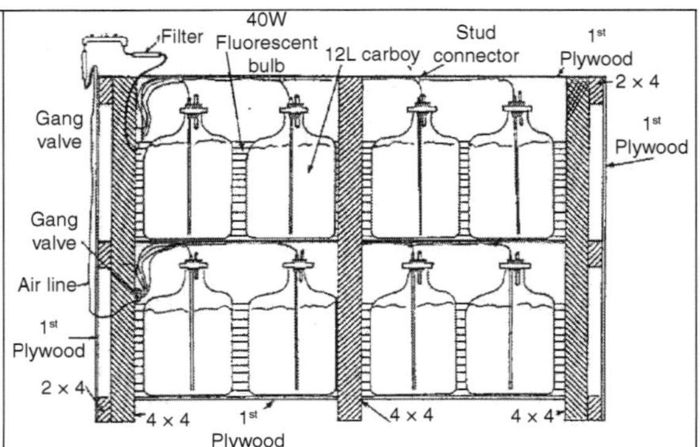

Carboy Culture Shelf

Batch culture systems are widely applied because of their simplicity and flexibility, allowing to change species and to remedy defects in the system rapidly. Although often considered as the most reliable method, batch culture is not necessarily the most efficient method. Batch cultures are harvested just prior to the initiation of the stationary phase and must thus always be maintained for a substantial period of time past the maximum specific growth rate. Also, the quality of the harvested cells may be less predictable than that in continuous systems and for example vary with the timing of the harvest (time of the day, exact growth phase).

Another disadvantage is the need to prevent contamination during the initial inoculation and early growth period. Because the density of the desired phytoplankton is low and the concentration of nutrients is high, any contaminant with a faster growth rate is capable of outgrowing the culture. Batch cultures also require a lot of labour to harvest, clean, sterilize, refill, and inoculate the containers.

Continuous Culture

Continuous (chemostat) - medium pumped into algal culture at slow, continuous rate

Advantages

Cultures can remain in exponential growth for long periods

Can control growth rate (must match dilution rate once cells are at equilibrium)

Disadvantage

Requires more expensive equipment (e.g., pumps)

Requires daily checking

The continuous culture method, (i.e. a culture in which a supply of fertilized seawater is continuously pumped into a growth chamber and the excess culture is simultaneously washed out), permits the maintenance of cultures very close to the maximum growth rate. Two categories of continuous cultures can be distinguished:

Turbidostat culture, in which the algal concentration is kept at a preset level by diluting the culture with fresh medium by means of an automatic system.

Chemostat culture, in which a flow of fresh medium is introduced into the culture at a steady, predetermined rate. The latter adds a limiting vital nutrient (*e.g.* nitrate) at a fixed rate and in this way the growth rate and not the cell density is kept constant.

Laing (1991) described the construction and operation of a 40 l continuous system suitable for the culture of flagellates, *e.g. Tetraselmis suecica* and *Isochrysis galbana*

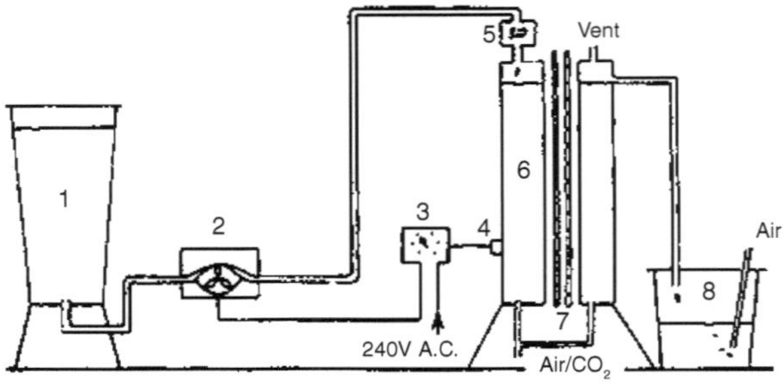

Diagram of a continuous culture apparatus (not drawn to scale):

(1) enriched seawater medium reservoir (200 l); (2) peristaltic pump; (3) resistance sensing relay (50- 5000 ohm); (4) lightdependent resistor (ORP 12); (5) cartridge filter (0.45 µm); (6) culture vessel (40 l); (7) six 80 W fluorescent tubes (Laing, 1991).

The culture vessels consist of internally-illuminatd polyethylene tubing supported by a metal framework. This turbidostat system produces 30 - 40 l per day at cell densities giving optimal yield for each flagellate species

Continuous culture methods for various types of algae in 40 l internally-illuminated vessels (suitable for flagellates only) (Modified from Laing, 1991),

Algae	Culture density for highest yield(cells per µl)	Usual life of culture(weeks)
Tetraselmis suecica	2 000	3-6
Chroomonas salina	3 000	2-3
Dunaliella tertiolecta	4 000	3-4
Isochrysis galbana		
Monochrysis lutheri		
Pseudoisochrysis paradoxa	20 000	2-3

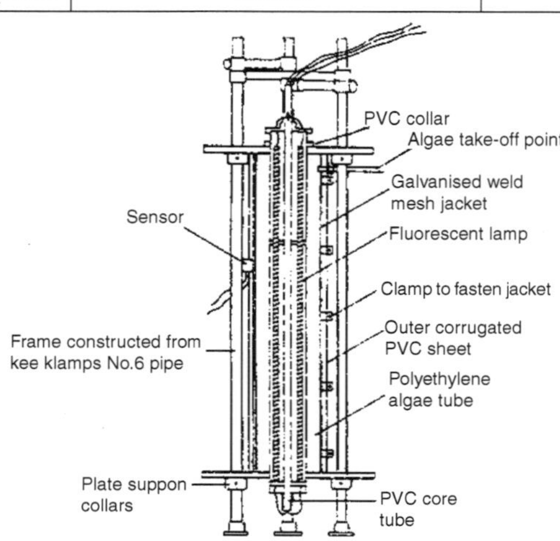

Schematic diagram (40 l continuous culture vessel)

A chemostat system that is relatively easy and cheap to construct is utilized by Seasalter Shellfish Co. Ltd, UK

Continuous culture of micro-algae in plastic bags.
Detail (right) shows inflow of pasteurized fertilized seawater and outflow of culture

The latter employ vertical 400 l capacity polyethylene bags supported by a frame to grow *Pavlova lutheri, Isochrysis galbana, Tetraselmis suecica, Phaeodactylum tricornutum, Dunaliella tertiolecta, Skeletonema costatum.* One drawback of the system is the large diameter of the bags (60 cm) which results in self shading and hence relatively low algal densities.

The disadvantages of the continuous system are its relatively high cost and complexity. The requirements for constant illumination and temperature mostly restrict continuous systems to indoors and this is only feasible for relatively small production scales. However, continuous cultures have the advantage of producing algae of more predictable quality. Furthermore, they are amenable to technological control and automation, which in turn increases the reliability of the system and reduces the need for labor.

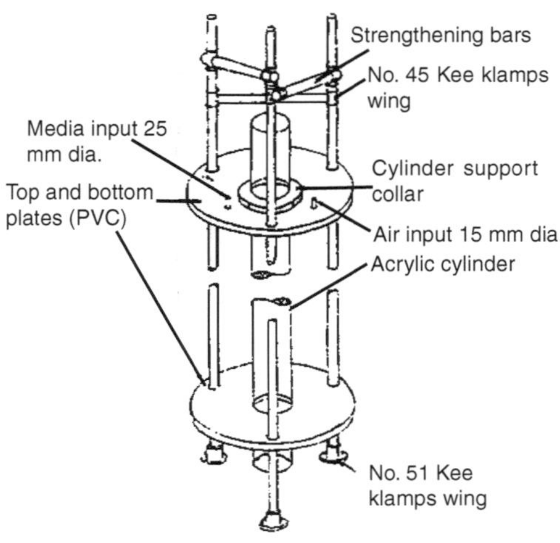

Continuous culture of micro-algae in plastic bags.

Detail (right) shows inflow of pasteurized fertilized seawater and outflow of culture

Semi-Continuous Culture

The semi-continuous technique prolongs the use of large tank cultures by partial periodic harvesting followed immediately by topping up to the original volume and supplementing with nutrients to achieve the original level of enrichment. The culture is grown up again, partially harvested, etc. Semi-continuous cultures may be indoors or outdoors, but usually their duration is unpredictable. Competitors, predators and/or contaminants and metabolites eventually build up, rendering the culture unsuitable for further use. Since the culture is not harvested completely, the semi-continuous method yields more algae than the batch method for a given tank size. In semi-continuous, remove and add a percentage of the medium at known intervals. The faster the interval, the more this approximates continuous culture.

Advantages - Cheaper than chemostats

Disadvantages - Balanced growth occurs only during exponential growth (in chemostats is the standard condition) so cell condition will change between fresh medium additions. Need to keep good track of fluorescence and other parameters to make sure you are in exponential growth phase

Algal Production in Outdoor Ponds

Large outdoor ponds either with a natural bottom or lined with cement, polyethylene or PVC sheets have been used successfully for algal production. The nutrient medium for outdoor cultures is based on that used indoors, but agricultural-grade fertilizers are used instead of laboratory-grade reagents. However, fertilization of mass algal cultures in estuarine ponds and closed lagoons used for bivalve nurseries was not found to be desirable since fertilizers were expensive and it induced fluctuating algal blooms, consisting of production peaks followed by total algal crashes. By contrast, natural blooms are maintained at a reasonable cell density throughout the year and the ponds are flushed with oceanic water whenever necessary. Culture depths are typically 0.25 - 1 m. Cultures from indoor production may serve as inoculum for monospecific cultures. Alternatively, a phytoplankton bloom may be induced in seawater from which all zooplankton has been removed by sand filtration. Algal production in outdoor ponds is relatively inexpensive, but is only suitable for a few, fast growing species due to problems with contamination by predators, parasites and "weed" species of algae. Furthermore, outdoor production is often characterized by a poor batch to batch consistency and unpredictable culture crashes caused by changes in weather, sunlight or water quality.

Mass algal cultures in outdoor ponds are commonly applied in Taiwanese shrimp hatcheries where *Skeletonema costatum* is produced successfully in rectangular outdoor concrete ponds of 10 - 40 tons of water volume and a water depth of 1.5 - 2 m.

Culture of Sessile Micro-Algae

Farmers of abalone (*Haliotis* sp.) have developed special techniques to provide food for the juvenile stages which feed in nature by scraping coralline algae and slime off the surface of rocks using their radulae. In culture operations, sessile micro-algae are grown on plates of corrugated roofing plastic, which serve as a substrate for the settlement of abalone larvae. After metamorphosis, the spat graze on the micro-algae until they become large enough to feed on macro-algae. The most common species of micro-algae used on the feeder plates are pennate diatoms (*e.g. Nitzchia, Navicula*). The plates are inoculated by placing them in a current of sand filtered seawater. Depending on local conditions, the micro-algae cultures on the plates take between one and three weeks to grow to a density suitable for settling of the larvae. As the spat grow, their consumption rate increases and becomes greater than the natural production of the micro-algae. At this stage, the animals are too fragile to be transferred to another plate and algal growth may be enhanced by increasing illumination intensity and/ or by the addition of fertilizer.

Chapter 10

Methods of Isolation of Micro algae

The isolation of an algal unit into a medium suitable for growth is required to establish a unialgal culture. There are several methods of isolation, i.e capillary pipette, agar plating, cover glass attachment, repeated subculture etc. The choice depends on the algal type and size. The method described below is applicable to many species. The source of algae for isolation is water sample containing mixed plankton population. This is suitable method for small species (<10 mm) or algae that grow well on substrate.

Agar Plating And Inoculation Techniques (Agar Plate Method)

Objectives

To demonstrate algal isolation and inoculation techniques.

To produce unialgal cultures from mixed plankton sample using agar plate method.

Materials

Microscope, Pasteur pipettes and bulbs, Erlenmeyer flasks, slides, cover slips, test tubes, Petri dishes, cotton plugs, alcohol lamp, wire loop, test tube rack, Walne's medium, F medium, agar, water sample containing mixed plankton population.

Procedures

All media, glassware's and other materials should be sterile.

a. Media Preparation

- Prepare agar plates and agar slants containing Walne's and F growth media separately, solidified with 1.5–2 % agar.
- Prepare liquid media in flasks using Walne's and F formulations.

- First add agar powder to medium at 1.5% concentration (1.5 g / 100ml medium)
- Then heat the medium to dissolve the agar, and pour it into Petri dishes.
- This agar solution will solidify as it cools.
- Plankton water should be diluted to several trial concentrations.
- Put a drop from each concentration of plankton water on a separate agar plate, and spread it over the surface of the medium with a glass rod.
- Then incubate the agar plates under favourable conditions.
- (Agar plates should be placed upside down so that water drops will not form on the lid and then fall on the culture).
- After 10 days or more, a few colonies of microalgae will appear on the surface of each plate.
- If one or more of the trial plankton concentrations is favourable, some of the colonies on at least one of the agar plates should be mono specific.
- Under a microscopic, pick up each mono specific colony together with some of the agar using a fine hook or other tool, if necessary.
- To determine if a colony is mono specific, it may be observed under a higher power microscope with a small amount of water added and transferred to an Erlenmeyer flask containing the same liquid medium to continue the subculture in a liquid medium.
- A sterilized tool should be used for each colony.

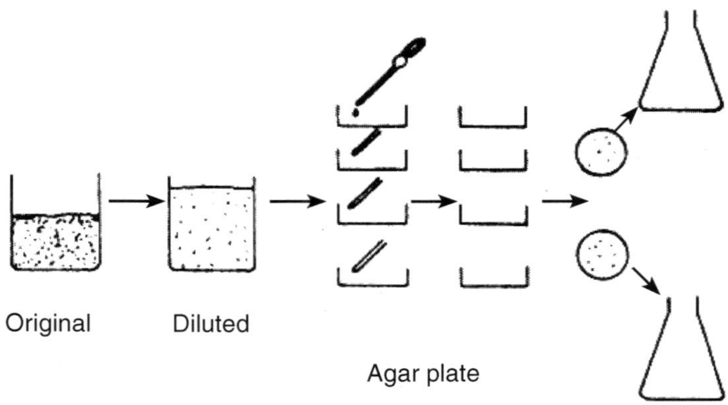

Agar plate Method

Streak Plating & Inoculation Techniques (Streak Plating Method)

This method of isolation is recommended when size of desired algal species is about 10 micron or less.

- Prepare Petri dishes containing sterile growth medium solidified by 1 - 1.5% agar

- Place 1- 2 drops of natural collection near the periphery of the agar.
- Flame sterilizes a wire loop or a glass rod using 70 % ethanol & an alcohol lamp.
- Make parallel streaks of the suspension on the agar cover, incubate the plate 4 - 8 days under suitable conditions for growth
- Through stereomicroscope, observe & select desired colonies for further isolation
- Remove a sample of the desired colony using a wire loop and place in a drop of sterile medium on a cover glass.
- Observe under high power objective of a compound microscope.
- Repeat streaking procedure using samples from a single colony
- Once growth is observed, transfer desired algal units to liquid or agar medium

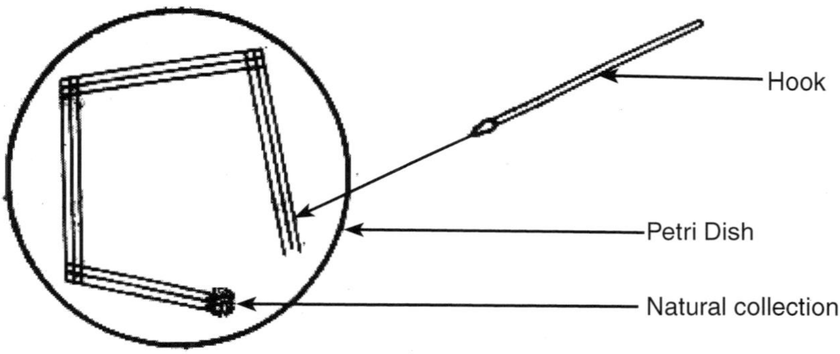

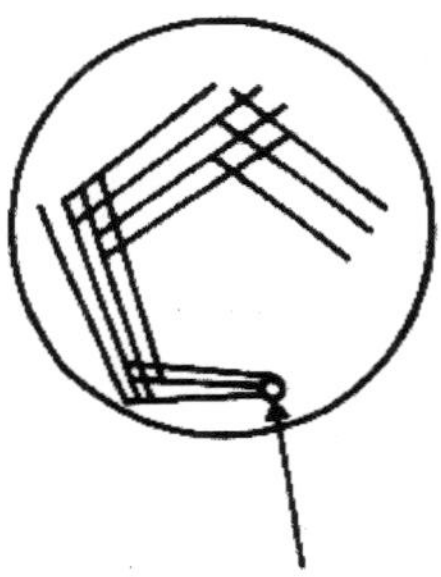

Pattern of streaks on agar surface; starting point is shown on tip of arrow

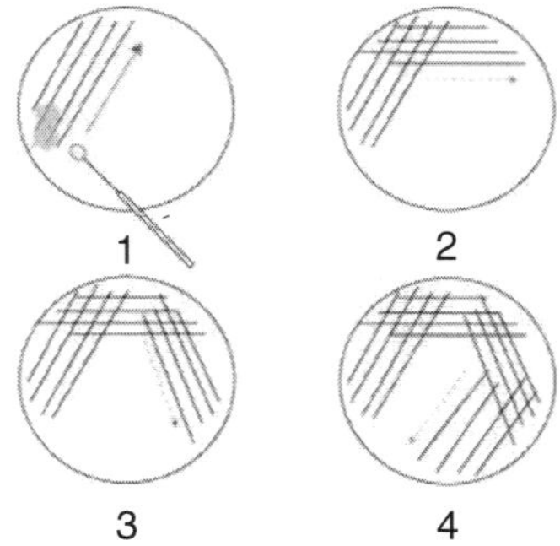

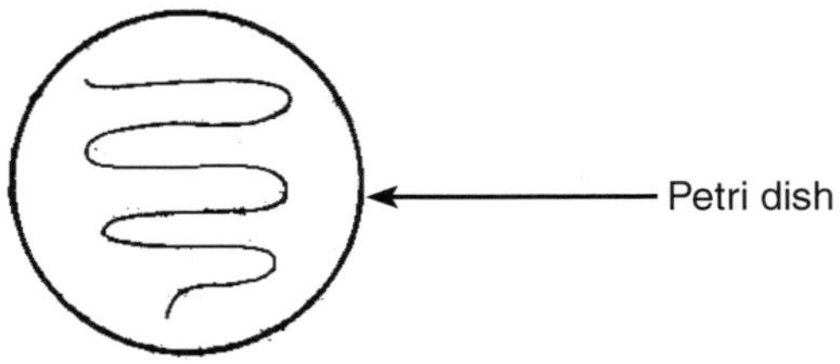

Pattern for streaking on agar

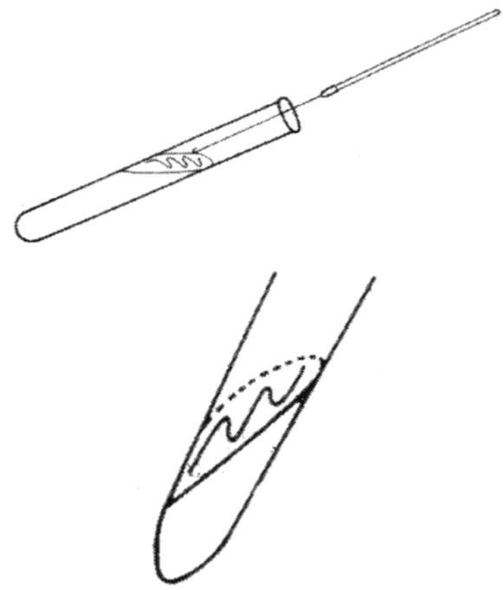

Pattern of streak on agar slant

Isolation on Agar

Separation of desired algal units from contaminants can be attained by inoculating them on an agar (solid) surface or by allowing the units to isolate themselves by creeping away from the contaminants.

- Prepare agar Petri dishes
- Remove the lids from the Petri dishes to allow the agar surface to dry
- Place a drop of raw material near the centre of the agar surface
- While observing under a stereomicroscope, manipulated single algal unit in a zigzag manner using a flame sterilized fine needle until isolation is achieved
- Draw the isolated unit using a capillary pipette or separate the unit by cutting a small agar block with a sterile micro spatula
- Transfer the isolated algal unit to a tube of liquid growth medium

Hooking and Pipetting Method

- For this method, pour the plankton water, dilute if necessary into a Petri dish.
- Then pick up the desired species with a fine hook or a capillary pipette under a microscope, preferably a binocular microscope.
- Transfer the organism into an agar plate and examine it again under the microscope to be sure only one species is present.
- If only one species is present, it may be incubated on this agar plate or transferred to liquid medium in a flask.
- If isolation is necessary to achieve a mono specific culture, either the repeated subculture method or agar plate method may be used or the hooking and pipetting procedure may be repeated.

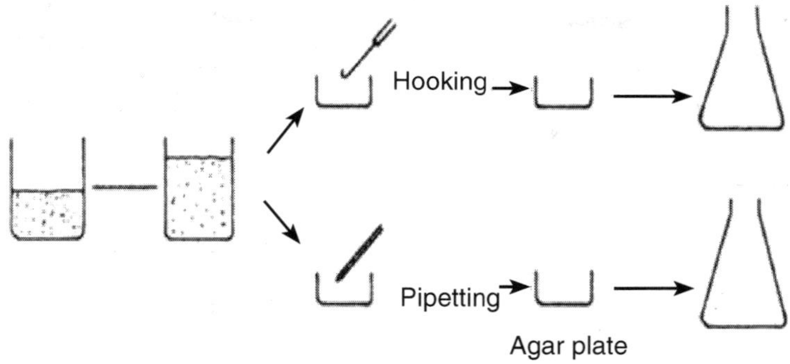

Hooking & Pipetting Method

Repeated Sub Cultures

- This is a very simple and very useful isolation method.
- Put a drop of the plankton water that was collected by plankton net or other means into each of several small Erlenmeyer flasks (50 to 100 ml) or test tubes.
- It may be necessary to dilute the plankton water to lower the concentration of organisms and reducing the number and kinds of organisms in a drop.
- The flask or test tube should contain a variety of media such as natural seawater, enriched water or artificial media of different salinities.
- Then incubate under different conditions of temperature, light intensities, illumination time, etc. After a few days under these conditions, a dominant species will appear in each flask.
- When this happens, repeat the process by transferring a drop from each flask to a new flask containing the same media and then incubate this new flask under the same conditions.
- This second isolation will probably produce mono specific culture.

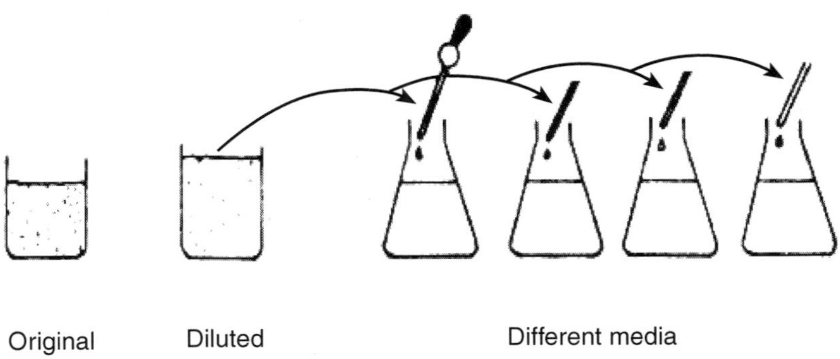

Repeated subculture

Serial Dilution Culture Technique

Equipment

- ➢ Culture tubes (sterile) screw-capped or sterile stoppered
- ➢ Test-tube racks, open mesh.
- ➢ Media — usually dilute e.g. f10, f20.
- ➢ Automatic dispenser (sterile) or 10 ml sterile glass pipettes.
- ➢ Glass pipettes 1 ml or pasteur (sterile), rubber or silicone teat.
- ➢ Bunsen burner or small flame.

Method

Using aseptic technique, dispense 9 ml of media into each of ten test tubes with Sterile automatic dispenser or sterile 10 ml pipettes. Label tubes 10-1 to 10-10 indicating dilution factor.

Aseptically add 1 ml of enrichment sample to the first tube (10^{-1}) and mix gently.

Take 1 ml of this dilution and add to the next tube (10^{-2}), mix gently.

Repeat this procedure for the remaining tubes ($1O^{-3}$ to 10^{-10}).

Incubate test-tubes under controlled temperature and light conditions:

a. Temperature and photoperiod — as close to the natural environment as possible.

b. Light intensity — slightly lower than the natural environment.

Examine cultures microscopically after 2 - 4 weeks by withdrawing small sample aseptically from each dilution tube. A unialgal culture may grow in one of the higher dilution tubes e.g. 10-6 to 10-10. If tubes contain two or three different species then micromanipulation can be used to obtain unialgal cultures.

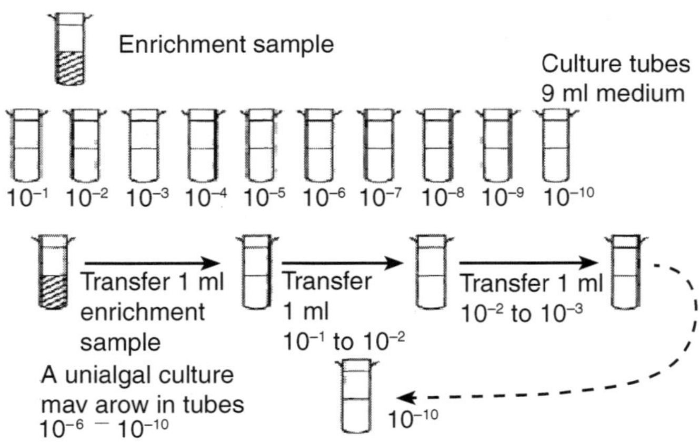

Micromanipulation

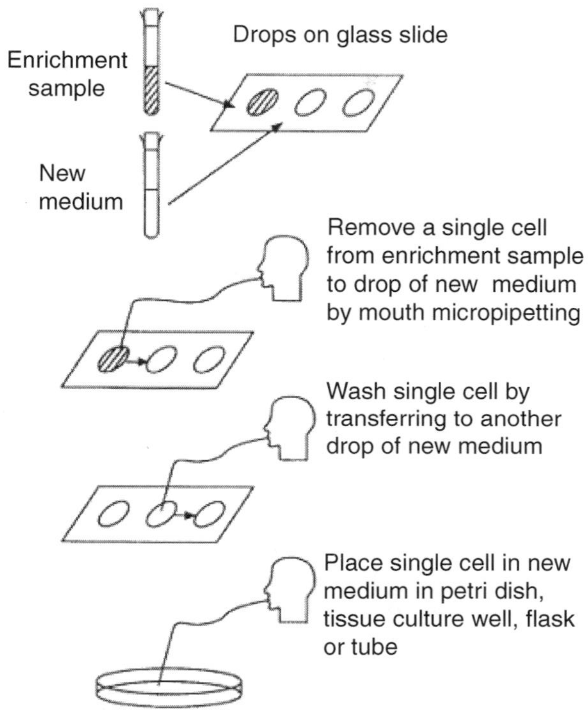

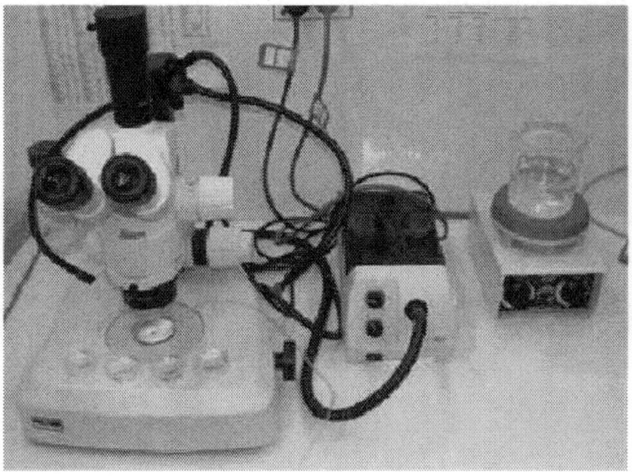

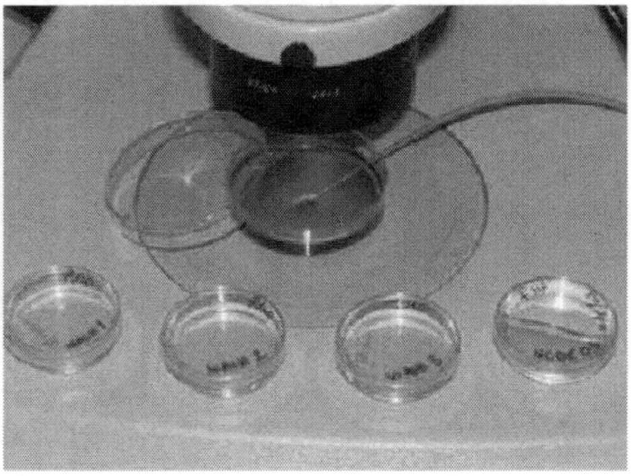

Equipment

Inverted microscope or stereo microscope with magnification up to 200 x, although 40 - 100 x should be sufficient in most cases. Phase contrast or dark field optics is an advantage.

Capillary tubes or haematocrit tubes - approx. 1 mm diameter x 100 m long.

Bunsen burner or small flame.

Silicone tubing to fit over end of capillary tube.

Length approximately 300 - 400 mm

Hot plate with beaker containing distilled water / Clean Glass slides

Agar plates (1.5 % Bacto-Agar, eg Difco Cat. No. 0140—01) made up in petri dishes (disposable, 90 mm dia) Tissue culture plates.

Pasteur pipettes (sterile), rubber or silicon teat.

Sterile media, usually at dilute nutrient concentrations; e.g. f_{20}, f_{50}

Sterile tissue culture multi-well plates or sterile disposable petri dishes

(e.g. 33 or 55 mm dia or sterile culture tubes).

Method

With a fine flames from a bunsen burner heat and draw out (holding at both ends) the capillary tube to form two micropipettes. The narrow end should be about twice the diameter of the cell to be micromanipulated.

Heat distilled water to simmering point on hot plate. This is used for sterilizing the micropipette between each transfer.

Place drops of sterile medium onto 1.5 % agar plates with a sterile pasteur pipette. Alternatively place three drops on a glass slide.

With silicone tubing attached to micropipette suck up and blow out with mouth a

small amount of hot distilled water. This sterilizes the micropipette.

Locate algal cell to be isolated in drop of enrichment sample. While observing the cell, suck up into the micropipette.

Transfer the cell to a drop of sterile medium on agar plate or glass slide.

Sterilize the micropipette.

Repeat this process to "wash" the cell. The more times a cell is washed the less likely is bacterial contamination. However, the risk of cell damage increases with the number of times a cell is handled. The optimum number of washes will depend on the type of algae.

Transfer cell to dilute medium in a tissue culture plate, petri dish or culture tube.

Place culture vessel under low light at appropriate constant temperature. Check

microscopically for growth or wait until macroscopic growth can be detected (3 - 4 weeks after transfer).

A clonal uni-algal culture should result from this method.

For large non-motile cells or when time is at a premium an alternative method is to isolate into petri dishes rather than on a slide. Multiple cells of a population are quickly transferred to the first rinse plate and then from there progressively fewer into subsequent plates. Because relatively less care is taken to avoid non-target cell transfer each petri dish acts as an enrichment culture with hopefully a relatively clean culture in the final plate (or even possibly clonal). The "clean" plates can then be used as the stock for making clonal isolations when time permits.

Capillary Pipette Method or Pipette Method

Large organisms can be pipetted out using a micro-pipette under microscope and transferred to culture tubes which are having enriched culture media. This method uses an inverted Petri dish top as an isolation dish.

- Place 10 - 15 drops of the natural collection in the centre of the dish
- Place 6 - 8 drops of suitable liquid medium in six positions encircling the natural collection. Each droplet is then assigned a numerical code
- With the use of sterile capillary pipette, transfer the desired algal units from the natural collection to one of the six drops. Desired algal units are located while looking through an inverted microscope or a stereomicroscope
- Transfer a single algal unit from the first drop to the second drop
- Repeat the process (moving clockwise) until a single algal unit is present in a drop of liquid medium
- Transfer the single algal unit to a sterile tube containing liquid culture medium.

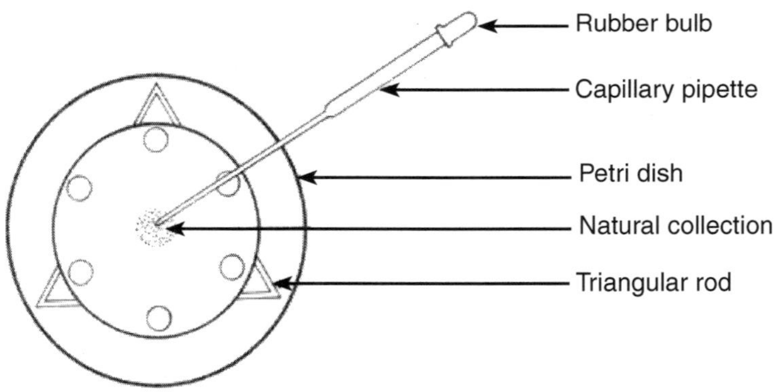

Inverted petri dish top

Centrifuge Method

- By repeated centrifuging of the sample and by inoculating the deposits, we may get different organisms.

Phototactic Method

- By this method, most of the phytoflagellates can be isolated.
- Make a dark chamber with a small hole on one side
- Keep a candle outside, near the hole

- Pour the water sample in a beaker and keep it inside the dark chamber, with its beak nearer to the hole.
- Since the flagellates have a tendency to move towards the light, they crowd near to the candle light.
- By pipetting, we can separate these organisms and a pure culture can be raised adopting culture tube method.

Chapter 11

Physical and Chemical Conditions for Algal Culture

The most important parameters regulating algal growth are nutrient quantity and quality, light, pH, turbulence, salinity and temperature. The most optimal parameters as well as the tolerated ranges are species specific and broad generalizations for the most important parameters are given below:

A generalized set of conditions for culturing micro-algae

Temperature (°C)	16 - 27 (18 – 24)
Light intensity (lux) (depends on volume and density)	1,000 - 10,000 (2,500 - 5,000)
Salinity (g / l or ppt)	12 - 40 (20 – 24) ppt
Photoperiod (light:dark, hours)	16 : 8 (minimum) / 24 : 0 (maximum)
Aeration	50 to 100% of the culture volume per minute
carbon dioxide	2 % of air volume
pH	7 - 9 (8.2 - 8.7)

Also, the various factors may be interdependent and a parameter that is optimal for one set of conditions is not necessarily optimal for another.

Culture Medium/ Nutrients

Concentrations of cells in phytoplankton cultures are generally higher than those found in nature. Algal cultures must therefore be enriched with nutrients to make up for the deficiencies in the seawater. Macronutrients include nitrate, phosphate (in an approximate ratio of 6:1) and silicate.

Light

As with all plants, micro-algae photosynthesize, *i.e.* they assimilate inorganic carbon for conversion into organic matter. Light is the source of energy which drives this reaction and in this regard intensity, spectral quality and photoperiod need to be considered. Light intensity plays an important role, but the requirements vary greatly with the culture depth and the density of the algal culture: at higher depths and cell concentrations the light intensity must be increased to penetrate through the culture (*e.g.* 1,000 lux is suitable for erlenmeyer flasks, 5,000 - 10,000 is required for larger volumes). Light may be natural or supplied by fluorescent tubes. Too high light intensity (*e.g.* direct sun light, small container close to artificial light) may result in photo-inhibition. Also, overheating due to both natural and artificial illumination should be avoided. Fluorescent tubes emitting either in the blue or the red light spectrum should be preferred as these are the most active portions of the light spectrum for photosynthesis. The duration of artificial illumination should be minimum 18 h of light per day, although cultivated phytoplankton develops normally under constant illumination.

Water pH

The pH range for most cultured algal species is between 7 and 9, with the optimum range being 8.2 - 8.7. Complete culture collapse due to the disruption of many cellular processes can result from a failure to maintain an acceptable pH. The latter is accomplished by aerating the culture. In the case of high-density algal culture, the addition of carbon dioxide allows to correct for increased pH, which may reach limiting values of up to pH 9 during algal growth.

Aeration/ Mixing

Mixing is necessary to prevent sedimentation of the algae, to ensure that all cells of the population are equally exposed to the light and nutrients, to avoid thermal stratification (*e.g.* in outdoor cultures) and to improve gas exchange between the culture medium and the air. The latter is of primary importance as the air contains the carbon source for photosynthesis in the form of carbon dioxide. For very dense cultures, the CO_2 originating from the air (containing 0.03 % CO_2) bubbled through the culture is limiting the algal growth and pure carbon dioxide may be supplemented to the air supply (*e.g.* at a rate of 1% of the volume of air). CO_2 addition furthermore buffers the water against pH changes as a result of the CO_2 / HCO_3 - balance. Depending on the scale of the culture system, mixing is achieved by stirring daily by hand (test tubes, erlenmeyers), aerating (bags, tanks), or using paddle wheels and jetpumps (ponds). However, it should be noted that not all algal species can tolerate vigorous mixing.

Temperature

The optimal temperature for phytoplankton cultures is generally between 20 and 24°C, although this may vary with the composition of the culture medium, the species and strain cultured. Most commonly cultured species of micro-algae tolerate temperatures between 16 and 27°C. Temperatures lower than 16°C will slow down growth, whereas those higher than 35°C are lethal for a number of species. If necessary, algal cultures can be cooled by a flow of cold water over the surface of the culture vessel or by controlling the air temperature with refrigerated air- conditioning units.

Salinity

Marine phytoplanktons are extremely tolerant to changes in salinity. Most species grow best at a salinity that is slightly lower than that of their native habitat, which is obtained by diluting sea water with tap water. Salinities of 20 - 24 g / l have been found to be optimal.

Carbon Dioxide

As its normal content in air is low (approx. 0.03 %), carbon dioxide is often supplied at 2 % by volume to optimize culture growth. Commercial grade CO_2 bottles are utilised, and the gas is injected into the main air pipe via a dispensing vessel in which gas bubbling reveals the gas flow. Since carbon dioxide is heavier than air, to prevent stratification the pipe makes some ups and downs after the point where it is injected.

Depending on the algal species, the CO_2 supply, and the volume of inoculum, working cultures normally reach their log-phase in 5 to 7 days. At this point, cultures can be utilized either to start new cultures, to be fed to rotifers or to be used as "green water" in fish tanks..

The volume of the algal inoculum is usually 15 to 20 % of the new volume. Smaller or larger inocula could be used to decrease or increase growth rate. In culture up-scaling, the new vessels have to be inoculated with a sufficiently high microalgae density in order to ensure a rapid growth and to limit the risk of contamination with different algae or other micro-organisms (protozoa, nematodes, fungi, etc.).

Chapter 12

Growth dynamics of Algal Culture

The growth of an axenic culture of micro-algae is characterized by five phases

Lag or Induction Phase

This phase, during which little increase in cell density occurs, is relatively long when an algal culture is transferred from a plate to liquid culture. Cultures inoculated with exponentially growing algae have short lag phases, which can seriously reduce the time required for upscaling. The lag in growth is attributed to the physiological adaptation of the cell metabolism to growth, such as the increase of the levels of enzymes and metabolites involved in cell division and carbon fixation.

Exponential Phase

During the second phase, the cell density increases as a function of time t according to a logarithmic function:

Ct = C0.emt with Ct and C0 being the cell concentrations at time t and 0, respectively, m = specific growth rate.

The specific growth rate is mainly dependent on algal species, light intensity and temperature.

Phase of Declining Growth Rate

Cell division slows down when nutrients, light, pH, carbon dioxide or other physical and chemical factors begin to limit growth.

Stationary Phase

In the fourth stage the limiting factor and the growth rate are balanced, which results in a relatively constant cell density.

Death or "Crash" Phase

During the final stage, water quality deteriorates and nutrients are depleted to a level incapable of sustaining growth. Cell density decreases rapidly and the culture eventually collapses.

In practice, culture crashes can be caused by a variety of reasons including the depletion of a nutrient, oxygen deficiency, overheating, pH disturbance or contamination. The key to the success of algal production is maintaining all cultures in the exponential phase of growth. Moreover, the nutritional value of the produced algae is inferior once the culture is beyond phase 3 due to reduced digestibility, deficient composition, and possible production of toxic metabolites.

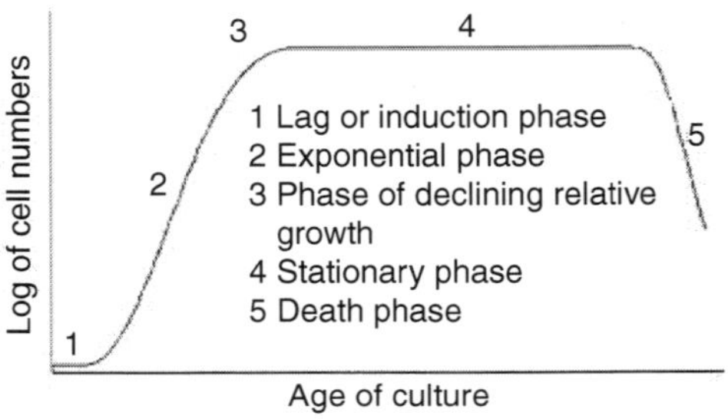

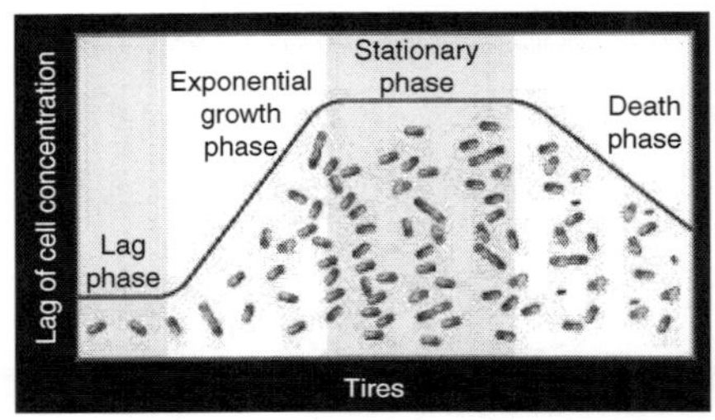

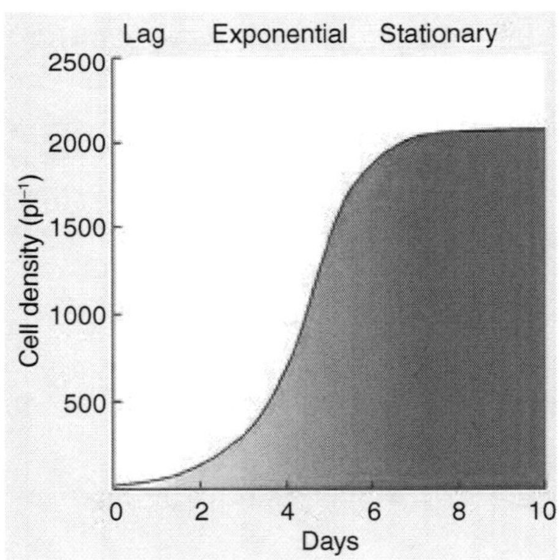

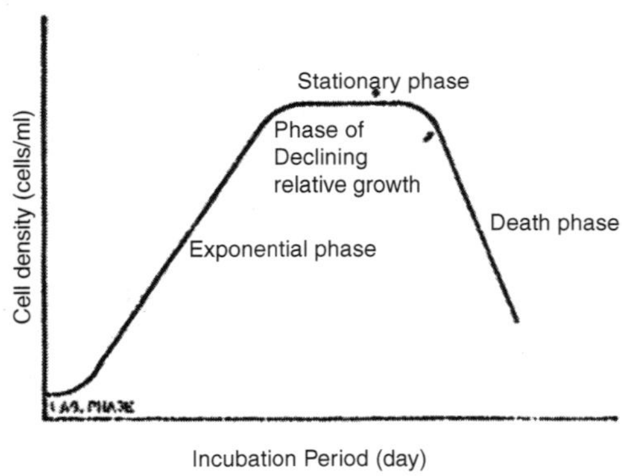

Chapter 13

Algal Culture Media

For culturing the micro algae, various culture media are in use depending on the organisms involved. Though Erd-Schreiber's and Miquel's media (Miquel, 1892) were found to be very effective for culturing the diatoms and nanoplankters, several other media are also available with the addition of trace metals, vitamins and other organic and inorganic salts. Since the diatoms require silica for building up the cell walls, the culture media should have a compound of silicate besides the nitrates, phosphates, chlorides and trace metals. Usually for culturing the flagellates, 'Conway' or 'Walne's medium (Walne, 1974) is used in the laboratory for the maintenance of the stock culture as well as mass culture. Since this culture medium has got various chemicals, trace metals and vitamins (B1 and B12), the phytoflagellates such as species of *Isochrysis, Povlova, Dicrateria, Chromulina* and *Tetraselmis* are being cultured by using media alone. However, media like TMRL and PM (Gopinathan, 1982) are also found to be effective for the mass culture of nanoplankton flagellates. Still, the technique of culturing different micro algae requires a clear understanding of their nutritional requirements, especially during the various phases of growth. The mportant culture media used for the micro algae are given below:

Schreiber's Medium

Potassium nitrate	0.1 g
Sodium orthophosphate	0.02 g
Soil extract	50 ml
Filtered and sterilized seawater	1 litre

Soil extract is prepared by boiling 1 kg of garden soil in 1 litre of freshwater for one hour. After 24 hrs, clear water is decanted and kept in a bottle. 50 ml of this soil extract can be added to each litre of sterilized seawater. This can be used as a medium while isolating the nanoplankton.

Miquel's Medium

Solution A	
Potassium nitrate	20.2 g
Distilled water	100 ml
Solution B	
Sodium orthophosphate	4 g
Calcium chloride	2 g
Ferric chloride	2 g
Hydrochloric acid	2 ml
Distilled water	100 ml

0.55 ml of 'A' and 0.50 ml of 'B' are added to one litre of filtered and sterilized seawater. This medium can be used for culturing various types of micro algae.

TMRL medium (Tung Kang Marine Research Lab)

Potassium nitrate	***10 g / 100 ml of DW***
Sodium orthophosphate	1 g
Ferric chloride	0.3 g
Sodium silicate	1 g

The chemicals are kept separately in 100 ml reagent bottle. 1 ml each to 1 litre of sterilized seawater is added. This medium can be used for the mass culture of diatoms.

'Conway' or Walne's Medium

Solution A	
Potassium nitrate	100 g
Sodium orthophosphate	20 g
Dist. water	1 litre
Solution B	
EDTA (Na)	45 g
Boric acid	33.4 g
Ferric chloride	1.3 g
Manganese chloride	0.36 g
Zinc chloride	4.2 g
Cobalt chloride	4.0 g
Copper sulphate	4.0 g
Ammonium molybdata	1.8 g
Dist. water	1 litre

Solution C (Vitamins primary stock)	
Vitamin B1 (Thiamin)	20 mg in100 ml distilled water
Vitamin B12 (Cyanocobalamine)	10 mg in100 ml distilled water

A, B and C (each) in different reagent bottles are prepared. 1 ml of A, 0.5 ml of B and 0.1 ml of C to 1 litre of filtered and sterilized seawater are added. During the course of the larval rearing of oysters, the flagellates form the basic food upto its spat stage. However, for the better growth of the spat, the food is to be changed from flagellates to a mixture of diatoms and other nanoplankters. For the preparation of mixture of various phytoplankton organisms in the open tanks, using direct sunlight, the following media could be used.

Mixture Culture Medium

Potassium nitrate	1.2 g
Sodium orthophosphate	0.66 g
EDTA (Na)	0.66 g
Sodium silicate	0.66 g

The first 3 chemicals are dissolved in 25 ml of DW and sodium silicate alone is dissolved in 25 ml of DW. This is added to 100 litre of fresh unfiltered seawater (fresh seawater is filtered through organdy cloth, 0.33 mm mesh size to remove zooplankton). This water is poured in 3 - 4 white lined basins or fibre-glass tanks and kept in open sunlight. Within 24 hrs, a slight yellow colouration can be noted. On examination, it can be noticed the growth of mixture of planktonic diatoms and other nanoplankters. If the temperature is very high and sunlight is very bright, only the blooming of *Chaetoceros* spp. could be observed.

Modified F Medium (modified from Guillard and Ryther)

Modified F Medium (modified from Guillard and Ryther)			
A.	***N–P Stock (500x)***		
	Sodium Ni rate	42.07	g
	Sodium Phosphate, monobasic	5.00	g
	Distilled water (to make)	1	litre
B.	***Sodium Metasilicate Stock (500x)***		
	Sodium Metasilicate	15.0	g
	Distilled water (to make)	1	litre
C.	***Ferric Chloride Stock (500x)***	***1.45***	***g***
	Distilled water (to make)	1	litre
D.	EDTA Stock (1 000x)		

Modified F Medium (modified from Guillard and Ryther)			
	EDTA, Disodium salt	10.00	g
	Distilled water (to make)	1	litre
E.	***Vitamin stock (1 000x)***		
	Thiamine HCl	0.2	g*
	B12	10.0	ml**
	Biotin	10.0	ml
	B12 Primary Stock *		
	- B12	0.1	g
	Distilled water	1	litre
	Biotin Primary Stock **		
	Biotin	0.1	g
	Distilled water	1	litre
F.	***Trace Metal Stock (1 000x)***		
	TM primary stock A	1	ml
	TM primary stock B	1	ml
	TM primary stock C	1	ml
	TM primary stock D	1	ml
	Distilled water (to make)	1	litre
1.	***TM Primary Stock A Solution***		
	Copper Sulphate, 5–hydrate	1.96	g
	Zinc Sulphate, 7–hydrate	4.40	g
	Distilled water (to make)	100	ml
2.	***Primary Stock B Solution***		
	Sodium Molybdate, 2–hydrate	1.26	g
	or		
	Ammonium Molybdate, 4–hydrate	0.907	g
	Distilled water (to make)	100	ml
3.	***TM Primary Stock C Solution***		
	Manganous Chloride	36.00	g
	Distilled water (to make)	100	ml
4.	***TM Primary Stock D Solution***		
	Cobalt Chloride	2.00	g
	Distilled water	100	ml

Besides the above mentioned laboratory prepared chemicals which serve as nutrients, commercial fertilizers can be used for the mass culture of diatoms and nanoplankters, in open tanks for economy purpose. The media used for the open culture are:

Urea 46	10 mg / l
16 - 20 - 0	10 mg / l
20 - 0 - 0	100 mg / l

Determination of Algal Cell Densities

Regular counts of the algal cells must be made in order to schedule inoculation of the various culture flasks as well as mass culture containers, to monitor growth of the algal cultures and to determine the quantity of algae to be fed to the rearing larval organisms. In sampling the mass or tank cultures, a small length of rubber tubing s helpful to work with and to connect it to the mouth piece end of a sterile serological pipette. To get a representative sample, move the pipette around the tank while withdrawing algae upto the mark on the pipette and then place it in a flask. This is how the sample is to be counted. Since most of the nanoplankters measure less than 10 μ, a haemocytometer is used for counting the cells. First of all the sample is treated with a drop of eosin or formalin to kill the cells and stirred well, one drop is taken with sterilized pipette. After placing a cover-slip on the haemocytometer, the pipette is brought to the edge of the haemocytometer to touch it. The sample runs inside the cover slip and thus a thin film of the culture is maintained, and the cells are equally distributed. Since the haemocytometer got 9 chambers and 4 sides, having 16 divisions and 5 chambers of multiple divisions, it is restricted the counting for at least 4 chambers. The average number of cells in 1 ml is calculated as shown below:

Average counts per chamber x 104 = Total number of cells / ml

Stock Culture Maintenance

Stock culture of the micro algae are maintained in a special air conditioned room adjacent to the mass culture room. Stock cultures are kept in 3 or 4 litre Haufkin culture flasks. The autoclaved or boiled seawater after cooling is poured to the Haufkin flasks and required nutrients are added. Walne's medium enriched with vitamins is the quite ideal one to maintain the stock of all the phytoflagellates. About 10 ml of the inoculum in the growing phase is transferred to the culture flasks and the same is placed in front of 2 tube lights (1000 lux). After 8 - 10 days, when the maximum exponential phase has reached, light is reduced to 1 tube light for further growth. The time required for the maximum cell densities varies depending on the species. However, it was noticed that all the Haptophycean flagellates require 2 weeks for the completion of growth phase before entering into the declining phase. In the stationary phase, the flagellate can be kept for 2 months in the stock culture room, under controlled conditions of light and temperature, with or without aeration. At the time of maximum exponential phase of growth, the colour of the culture turns into dark brown and the cells are found as suspension without movement. The cells of flagellates produce a thick matrix or cyst around it in the stationary phase and when inoculate the same, these cysts break and the flagellates eme rge out for its further growth and multiplication. A minimum of 5 Haufkin culture flasks were kept for each species as stock culture.

Mass Culture

Large scale culture of micro algae, especially nanoplankton flagellates and selected species of diatoms are necessary for feeding the larval forms in a hatchery. Since the molluscan larvae feed organisms measuring < 10μ only, these forms are isolated from the seawater, maintained as stock culture and the inoculum, mass culture is done in the laboratory conditions as well as in the outdoor tanks. The containers for the mass culture of micro algae are of 10 L capacity polythene bags, 20 L glass car buoys, 100 L Perspex tanks and 250 L cylindrical transparent FRP tanks for the indoor culture. These containers are kept in wooden racks with light and aeration. Fully grown culture from the stock culture room is used as inoculum for the mass culture in these containers. These tanks have the maximum concentration of the cells in the growing phase on the 5 - 7th days and harvested. After estimating the cell concentration using a haemocytometer, the culture is supplied to the hatchery for the rearing operations, of the larval organisms. Leaving 2 L of the culture, fresh enriched medium is added for further culture in the same container.

Harvest

The fully grown culture should be harvested during the exponential phase of the micro algae after determining the cell concentration. If the culture has entered the declining or stationary phase, the metabolites will be very high and the cells may not be in healthy condition. The rearing larval organisms may not show the expected growth if fed with this feed.

Preservation

The maintenance and constant supply of the culture whenever required is a problem in the hatchery especially during adverse weather conditions. In the case, the preservation of the algae either by freezing or by sun drying is done in the sense that during scarcity of the feed, the rearing operations may be successfully controlled. For the method of freezing, the culture is flocculated either by adding lime or by adjustment of pH using sodium hydroxide. Usually, the addition of alum or lime will not give satisfactory results, but manifestation of pH with sodium hydroxide is the advanced technique conducted in the laboratory. After knowing the quantity of the culture to be flocculated, the volume of sodium hydroxide solution needed to flocculate to get one degree raise in pH is measured. Suppose the pH of the culture is 8.4 by adding sufficient quantity of sodium hydroxide solution raised to 9.4. After vigorous stirring, the culture is left for one hour to settle the algal mass at the bottom. Slowly decanted the clear water and the mass in a plastic bucket is collected. Then the pH of the culture to the original level is brought slowly by adding dilute hydrochloric acid. Now the alga is ready for freezing or sun-drying. The algae are dried by pouring the mass in white enamel trays and keeping the same in the bright sun-light. If the algae dried thoroughly, the powder from the enamel tray is scraped and kept in glass bottles. Before freezing the algal mass, some protective reagents like Dimethyl sulphoxide ore Glycerol (a few drops) are added. The concentrate is then poured into polythene bags after measuring. The polythene bags ar e labeled and kept in deep freezers. The frozen algae do not have the same protein content as in the live condition. Whenever adverse condition arises, the frozen foods are used for rearing the larval organisms.

Determining Algal Densities

Monitoring algal growth in mass tank cultures is essential for successful production. Cultures that have stopped growing are either:

1. Significantly contaminated with microscopic predators and/or competitors, or
2. Have reached maximum ceil density under the present conditions.

A general gauge of algal growth is the daily darkening color as the cells multiply. A more accurate measure of growth can be made by comparing actual cell numbers with the previous day's count.

The number of cells can be determined by studying a sample from a culture under a 400X microscope. The sample is placed on a special slide called a hemacytometer. These slides are available from scientific or medical supply stores and come with instructions. To use the hemacytometer, place a pipet filled with a sample from the culture tank against the "V" groove of the slide (a cover slip must be in place). Withdraw the pipet before the sample runs off the flat grid area, but not before the grid area is completely covered with the algal suspension. Wait about five minutes for the cells to drift onto the grid. The grid area resembles the pound symbol (#) inside a square border so that there are nine equal squares. Each of the four corner squares are further divided into 16 smaller squares. The total numbers of algal cells in all 16 small squares in the four corner areas (64 small squares.) are counted. Divide this number by four, for an average, and multiply by 10,000 (hemacytometer volume factor). The product is equal to the number of cells per ml in the sample. Motile algae like *I. galbana*, must be killed before counting. A weak toxin such as acetic acid or ethanol may be used (one drop into 20 - 30 ml of algal sample). Dense cultures (>3 million) should be diluted 1:10 to facilitate counting and to reduce error. Remember to multiply by 10 (for a 10 : 1 dilution) or by the appropriate correction factor if a different dilution is made. There are numerous other methods to quantify algal densities, however using a microscope and hemacytometer, will reveal the presence of biological contaminants

Peak or maximum density (for existing conditions) is attained when, on consecutive days, the cell count does not increase by a factor of two. With practice, approximate cell counts can be estimated by observing the gross color of each culture. These estimates are based on the tint of the water caused by the algal cells and comparing it to the actual cell count. Tedious and time consuming cell counts (or other growth measurements) can be eliminated from the daily routine. Cell counts should be made periodically as a check on the gross estimate.

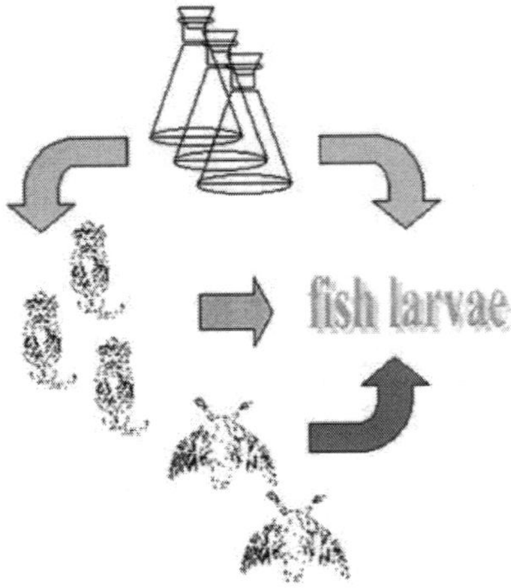

Simplified trophic chain established in the hatchery

Mass Production Systems

For aquaculture purposes, microalgae are mass produced in three main ways: (i) batch (or discontinuous or multistep back-up system) culture, (ii) semi-continuous culture and (iii) continuous culture.

Chapter 14

Biology and Culture Requirements of Rotifers

Taxonomy

PHYLUM	:	Nemathelminthes or Aschelminthes
CLASS	:	Rotatoria
ORDER	:	Monogononta
FAMILY	:	Brachionidae

Morphology

Rotifers are among the smallest filter-feeding metazoans. Composed of a fixed number of about 1,000 cells, their growth is obtained by plasma increase, not by cellular division. They filter small particles by means of a ciliated annular organ, the corona, located in the anterior part of the body. The corona is also used for its whirling locomotion, hence the name of the Class Rotatoria. Whereas many species spend their life span attached to a substrate by means of their retractile foot, *Brachionus plicatilis* that is the main species cultured for finfish larval rearing worldwide is a planktonic, unattached rotifer.

The epidermis contains a densely packed layer of keratin-like proteins called lorica. The rotifer's body is differentiated into three distinct parts: head, trunk and foot.

- The head carries the ring organ or corona, which is easily identifiable by its ring of cilia. The retractile corona assures locomotion and a whirling water movement which facilitates the ingestion of small food particles (mainly algae and detritus).
- The trunk contains the digestive tract, consisting in a mastax that grinds

the ingested particles, the oesophagus, the stomach with gastric glands and the intestine. The excretory system consists of paired protonephridia with terminal cells (cyrtocytes), the duct and the bladder. The protonephridia expel liquid excreta and play an important role in osmotic regulation. The genital organ is unpaired (*Monogononta*) or paired in the *Seisonidea* and *Bdelloidea*. The joint external opening for bladder and oviduct is called the cloaca.

- The foot is a ring-type retractile structure without segmentation, ending in one or four toes bearing pedal glands that secrete an adhesive substance in crawling and sessile rotifers.

Two different morphotypes of *B. plicatilis* exist: the small (S) type and the large (L) type. They differ in their lorica length: 130 to 340 µm (average 239 µm) for the L - type and 100 to 210 µm (average 160 µm) for the S - type. There are also differences in weight, shape of occipital spines and optimal growth temperatures (L - type rotifers have a wider temperature range while S - type rotifers have a higher temperature resistance). S - type rotifers are suitable as first food for fish larvae with a mouth opening smaller than 100 µm at first feeding, such as gilthead seabream, groupers and rabbitfish.

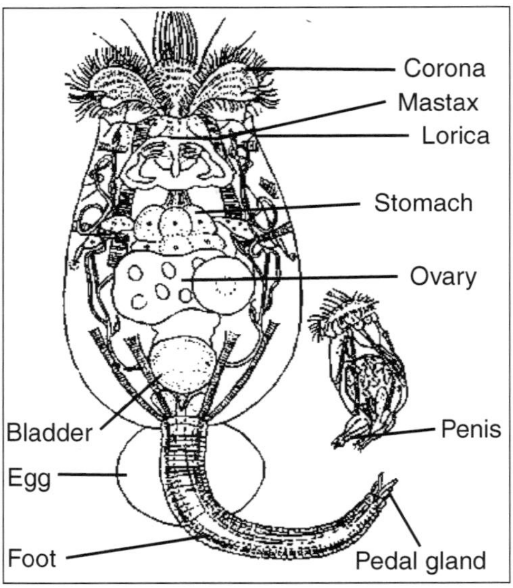

***Brachionus plicatilis*: female and male**

Two different morphotypes of *B. plicatilis* exist: the small (S) type and the large (L) type. They differ in their lorica length: 130 to 340 µm (average 239 µm) for the L - type and 100 to 210 µm (average 160 µm) for the S - type. There are also differences in weight, shape of occipital spines and optimal growth temperatures (L - type rotifers have a wider temperature range while S - type rotifers have a higher temperature resistance). S - type rotifers are suitable as first food for fish larvae with a mouth opening smaller than 100 µm at first feeding such as gilthead seabream, groupers and rabbitfish.

Life History

The life span of rotifers is measured in days and depends on culture temperature, but in a controlled environment and at 25°C, it has been estimated to range around 7 days. At this temperature, larvae become adults after 0.5 to 1.5 days and then females start to lay eggs approximately every four hours. It is believed that a female can produce ten generations of offspring. The reproductive activity of *Brachionus* is also influenced by temperature as illustrated below:

Effect of temperature on reproductive activity of *Brachionus plicatilis*(after Rutner-Kolisko 1972)

Temperature (°C)	15	20	25
Embryonic development (days)	1.3	1.0	0.6
First spawning of females (days)	3.0	1.9	1.3
Interval between spawning (hours)	7.0	5.3	4.0
Life span (days)	15	10	7
Number of eggs pre female during its lifetime	23	23	20

The reproduction of *B. plicatilis* can be either sexual (called mictic reproduction) or asexual (amictic or parthenogenetic reproduction). Only the latter is adopted for rotifer mass culture due to its faster rate and also due to the absence of males, which are useless as fish feed not having a functional digestive tract.

In the amictic reproduction the offspring are clones genetically identical to their mothers, i.e. all newly born rotifers are diploid females. Such multiplication can go on for months in a population kept in proper rearing conditions. Depending on environmental conditions, each female may produce about 20 amictic eggs.

Males are only produced after a sudden change in the environment, when females produce haploid (n chromosomes) eggs. Males and females breed, and the result is mictic resting eggs, analogous to *Artemia* cysts, which will hatch amictic females again.

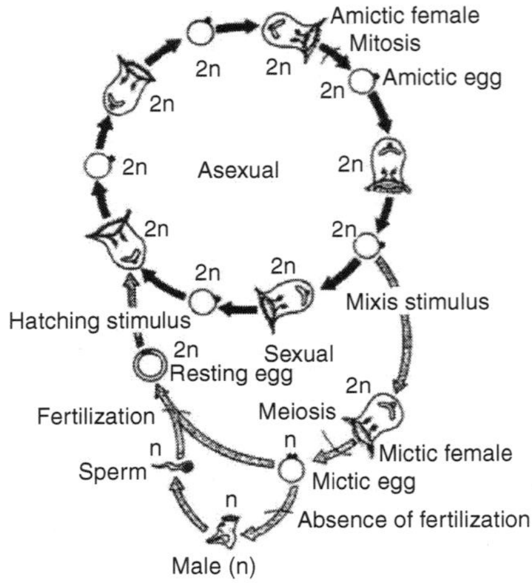

Food and Feeding

Chlorella has probably been the most popularly used algal food for the culture of *Brachionus plicatilis*. Use of *Chlorella, Dunaliella, Monochrysis lutheri, Nannochloris* sp., and *Dunaliella, Tetraselmis suecica* was a more suitable feed than *Chlorella* based on the high yields of *B. plicatilis* he obtained. The use of dried and ground blue green alga Spirulina maxima also proved promising.

Owing, however, to some difficulties encountered by many in mass producing sufficient supply of phytoplankton species for rotifer culture, the possibility of utilizing other types of food was looked into.

It was in 1967 when Hirata and Mori introduced the use of baker's yeast for rotifer culture. Further confirmation of the results they obtained led to a number of studies conducted on yeast in Japan. Other experimental diets were tried like Torula yeast and "food" yeast. In 1973, Hirayama and Watanabe also tested caked yeast, *Rhodotorula* sp., and commercialized dried marine yeast (ASY-4011). He concluded that yeast may be used most effectively as a supplemental food when phytoplankton supply is not sufficient. Another marine yeast species, *Zygosaccharomyces marinus*, which was isolated also found to be a good food for *Brachionus* and zoeal larvae of *P. japonicus*.

Mass-Culture of Rotifer (*B. plicatilis*) by Two Ways

In the first method, the feed for rotifer is added to the rotifer tank whenever it is required. In other words, feed preparation and rotifer rearing are separate aspects of the culture programme. Trotta and SEAPDEC follow this method.

Trotta has made use of a continuous system, in which Chlorella is grown with sterile culture medium in one large bag. This Chlorella-bag has separate interconnections to a reservoir, containing nutritive medium (for phyto-growth) and a rotifer-bag, which is twice the size of the Chlorella-bag, As soon as Chlorella has attained optimum concentration, it is allowed to flow from the Chlorella-bag into the bottom of the rotifer-bag, which expels a portion of rotifer medium through the outlet at the top of the rotifer-bag. Rotifer will be harvested from the medium, thus expelled. Subsequently, the Chlorella-bag will be refilled with fresh enriched medium, obtained from the reservoir, thereby making the above culture process as a continuous one. In SEAFDEC, 350 ton capacity concrete tanks are separately used for rotifer and Chlorella culture. Chlorella is cultured by fertilizing filtered seawater with inorganic fertilizers such as urea, superphosphate and ammonia. This Chlorella water is siphoned into the adjacent rotifer tank which is already inoculated with a stock culture of B, plicatilis. When the rotifer population attains a density of 200 - 300 numbers / ml, part of the rotifer medium is siphoned off through a fine-meshed phytoplankton net to harvest the rotifer. Water level' in rotifer tank is restored by replenishing with *Chlorella* water from the *Chlorell*a tank. After a few days, rotifer will again attain the harvestable density and once again the same process has to be repeated. Thus the whole proceiss goes on continuously until the culture becomes contaminated by unwanted organisms either in the *Chlorella* tank or in the rotifer tank.

In the second method, both rotifer and its feed are cultured together in one and the same container and this method is followed at Narakkal Prawn Culture Labratory.

At Narakkal Prawn Culture Laboratory, outdoor containers of 2 ton, 10 ton and 40 ton capacity are used for the mass culture of *B. plicatilis*. Initially, feed for rotifer is developed prior to rotifer inoculation in these tanks by fertilizing the filtered (through 50 micron net) seawater with groundnut oil cake (juice), urea and superphosphate at the rate of 200 g, 2 g and 2 g respectively per ton of water. Profound aeration is necessary and a starter culture of *Chlorella* is inoculated. After taking these measures to develop the rotifer feed, a starter culture of *B. plicatilis* is added to the same container when pH of the medium increases above 7. The rotifer starter should be obtained from a healthy population and it is preferable to be with egg-bearing parthenogenetic females. It must be devoid of either males or cyst bearing females. Inoculation is preferable to be at the rate of females. Inoculation is preferable to be at the rate of 5 - 10 rotifers per ml of the medium. A Chlorella bloom develops within 2 - 3 days and the rotifers multiply rapidly, attaining a population density of about 250 number / ml within 4 - 6 days after which harvesting is done everyday in the morning when rotifers swam at the surface. pH and oxygen content of the medium play an important role in rotifer multiplication, chluter (1930 a) has observed that at pH above 9.5 and below 6.5 rotifer is survived and high rotifer densities are associated with a pH value of 6 - 8. If dissolved oxygen content is above 1.15 mg O_2 / l, the reproduction rate is not inhibited while rotifers ceased to reproduce and die within a few days at 0.72 mg O_2 / l, The presence of a male or a cyst bearing female indicates the existence of adverse conditions in the culture tank. Hence routine observation is essential in order to understand the biological and physical conditions prevailing in the culture tank and the growth pattern of the population. Chlorella bloom declines, half the volume of water is replaced with fresh seawater and refertilized with the abovementioned fertilizers at half of the initial dose. The culture process is repeated until the culture gets contaminated with filamentous blue green algae or ciliates. If contamination occurs, the present culture operation has to be closed and the whole process has to be restarted from the beginning. When this culture method is followed, population may reach upto a high density of 560 rotifers / ml.

Culture Techniques

A. The "Daily-Tank-Transfer" Method

(Both rotifer and its feed are cultured together in one and the same container)

In this method, *Brachionus* is continuously subcultured using 0.5 m^3 tanks. The tanks are initially used for *Chlorella* cultures. When *Chlorella* density reaches about 10–20 × 10^6 cells / ml inoculation of rotifers is done. Once *Brachionus* yield increases significantly, the rotifers are harvested and transferred to another tank

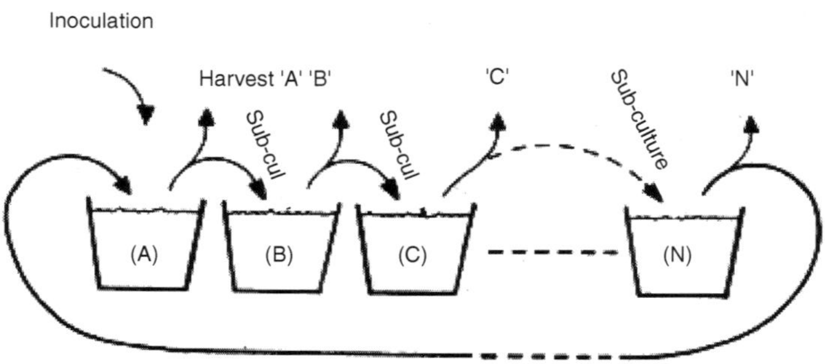

Schematic diagram of initial culture method for Rotifers, called daily tank-transfer

This method was developed in the Yashima station of the Seto inland sea farming fisheries association during 1964 - 1967. Several 0.5 m^3 slate tanks were used for producing both marine *Chlorella* and Rotifers. The process of transfer or subculturing continues for an indefinite period of time. The main disadvantage of this method is that it is too labour intensive.

B. The "Drain-Off" System for Large Tank Outdoor Culture

(both rotifer and its feed are cultured together in one and the same container)

This extensive method of rotifer culture was first developed in Japan using large concrete tanks. The procedure involves the following steps:

a. Culture marine *Chlorella* using modified Hirata medium until density reaches 10 - 20 × 10^6 cells / ml.

b. Fertilize again using double strength of above-mentioned culture medium.

c. When *Chlorella* reaches stationary phase of growth, *Brachionus* is inoculated at an initial density of 10 - 50 individuals / ml.

d. Once *Chlorella* is consumed, introduce additional food or baker's yeast (at 1 gallon / million rotifers / day).

e. When *Brachionus* density reaches 120 - 150 individuals / ml harvest about 20 - 30 % of the culture and transfer to another tank.

The procedure is thus repeated until the water becomes polluted with faecal materials. A diagram of the "drain-off" system is shown in Figure as given below:

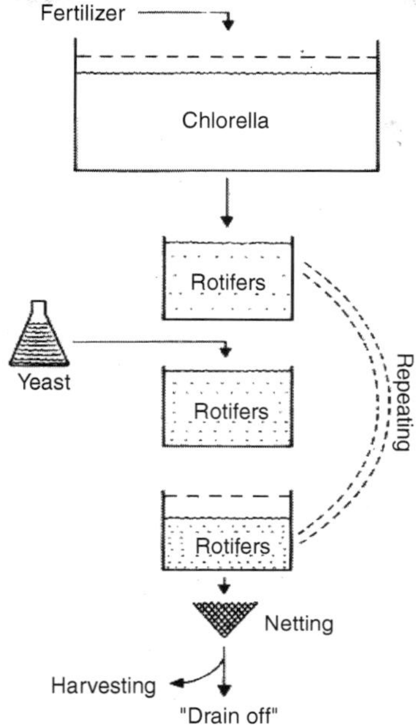

Mass culture of Rotifers (drain-off system in large outdoor tanks)

C. The "Feedback" Culture System

Hirata (1979) describes feedback as "feeding resulting in excretion, then excretion converted into feeding again". In an artificial culture system, "self-purification" which is carried out smoothly in the natural waters, cannot be attained due to the accumulation of biodeposits produced by the cultured organisms. Therefore, a system should be devised to promote energy flow and this is the very principle on which Hirata's feedback system was evolved. A flow chart of the principle of the feedback culture system is presented in Figure as given below:

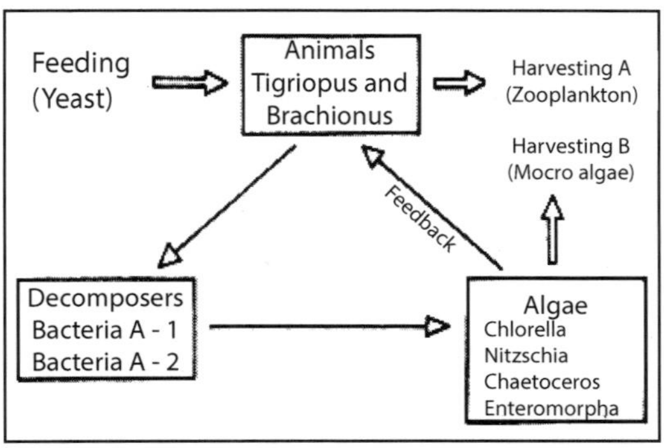

Flow chart of marine zooplankton culture in a feedback system

In a feedback system, marine zooplankters like *Tigriopus* and *Brachionus japonicus* (or the harpacticoid copepod, *Tisbintra elongata*) may be cultured together. These organisms serve as consumers. The microalga, *Chlorella* and the macroalga *Enteromorpha intestinalis* are initially cultured as producers. Bacteria which grow in the tank will then serve as decomposers.

There are two tanks in this system - a circular tank which serves as reservoir and a rectangular tank called the zigzag stream unit. The water in the first tank is recirculated to the stream by an airlift pump at the rate of 20 times a day. The macroalga, *Enteromorpha intestinalis* is grown together with the zooplankton in the zigzag stream. Framed nylon nettings are placed in each of the slots in the stream to serve as attachments for *Enteromorpha* which is filamentous. Water quality can be maintained for long periods of time in this type of culture system. While this shows a conceptual diagram of a feedback system for mass-scale production of rotifers.

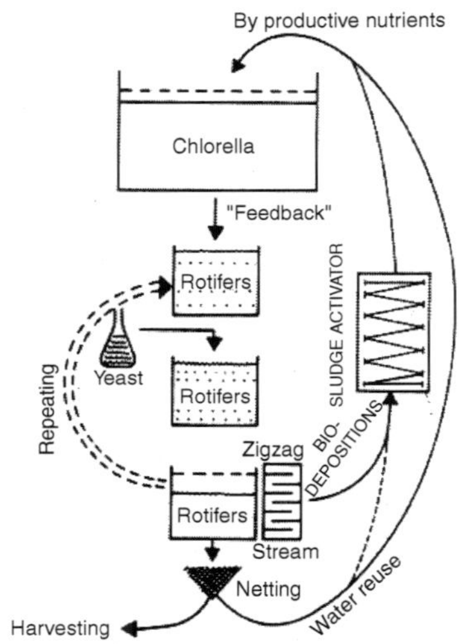

Diagram Mass culture of Rotifers (Feedback system)

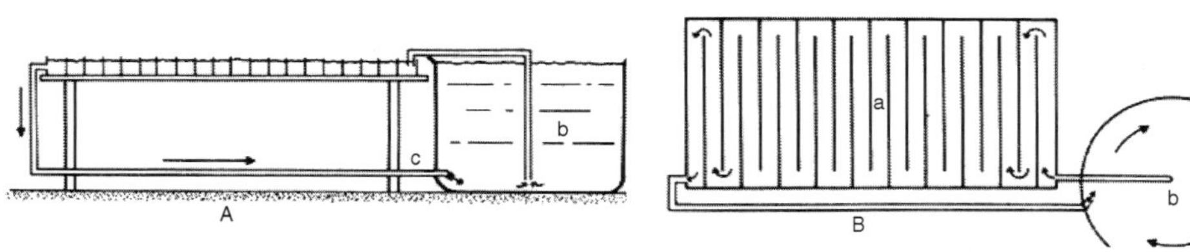

Schematic diagram of main tank and zigzag stream unit for cultivation of zooplankton (*Tigriopus* and *Brachionus*) and macro alga (*Enteromorpha*) A. Side view; B. Top view; a. 150 l zigzag stream unit; b. 550 l main cultivation tank; c. Recirculation pipe.

For outdoor cultures of *Brachionus*, one-ton fiberglass tanks are used and rotifers rely mainly on *Chlorella virginica* for food. The use of baker's yeast is also employed especially when *Chlorella* supply is not sufficient. The use of other types of food is presently being experimented on.

Tanks are regularly cleaned and disinfected using dilute (about 10 %) muriatic acid solution. Thorough rinsing is done prior to use. During rainy months, the tanks are usually provided with framed plastic sheets which serve as covers.Water utilized for culture purposes undergo filtration provided by improvised sand filters. Water salinity for rotifer culture is usually within the range of 25 - 28 ppt. Water temperature ranges between 27 - 30°C.

Antibacterial Treatment

As the rotifer may harbour bacteria such as *Vibrio, Pseeudomonas, Moraxella, Cytophaga* and *Flavo*bacterium, antibacterial treatment is required. Sodium nifurstyrenate (NFS-Na) at the rate of 5 microgram / ml with ultra violet radiation is most effective.

Tank Preparation

Before starting a new production cycle, normally after the harvest of the previous rotifer culture, prepare the tank as follows:

1. Rinse with tap water to eliminate the bulk of organic debris;
2. Wash it thoroughly with brush and detergent and rinse it again;
3. Wash or spray the tank walls with 500 ppm active chlorine solution; after a couple of hours, drain the tank and rinse it well until the chlorine smell is gone;
4. Let the tank dry and fill it with sterilized heated water only when needed; as an alternative possibility, fill the tank with seawater and sterilize with hypochlorite, then neutralize the residual chlorine with sodium thiosulphate

Repeat the procedure for the equipment to be used in the tanks: aeration tubing, drain valves and suspended traps. A practical procedure is to assemble all small equipment in the new tank, fill with SW and sterilize with hypochlorite: the equipment will be disinfected as a consequence.

To be used as inoculum, the rotifer population must still be in the middle of its log phase with at least 20 % fertility rate (measured as percentage of eggs over total rotifers). Never use rotifers which have already reached the last phase characterised by limited motility, scarce repletion and no egg presence.

In algae/yeast fed tanks, the initial density of inoculum, one of the most important factors in rotifer culture, should be kept at least at 100 animals / ml, with an optimal density of 150 - 200 animals / ml. High density cultures with artificial feeding need up to 500 rotifers per ml. With an initial density of 200 animals / ml, rotifer density should reach its peak within four to six days at 25°C.

Tank inoculum may come from large rotifer bags (vertical upscaling), or from other tanks (horizontal upscaling). One day before being harvested to be used as inoculum, rotifers in the tanks should be fed a fertility booster such as Protein Selco® by INVE SA. In order to reduce contamination and culture crash risks, horizontal upscaling should be limited to 7 - 8 cycles, after which the inoculum should be again taken from large bags.

Cleaning Cultures

In round tanks with conical bottom, bottom sediments are removed by letting them first settle for 10 - 15 minutes in absence of aeration (take care to check DO levels during the operation). Then the bottom valve is opened for a few seconds and it is closed when the outflowing water is again clean. Repeat the operation twice a day, in the morning and evening.

Complete the cleaning procedure by removing any greasy layer that may be formed at the water meniscus with a sponge or a paper cloth. Never dip your hands in water.

The rotifer tank water can be rapidly polluted by particulate matter, faeces and flocks of uneaten food. Their removal may prevent an excessive bacterial development and will increase the oxygen available for rotifers. As the continuous aeration of the water volume prevents in part the settlement of particles, their removal is achieved by means of "particle traps". These devices consist of floating mats of coarse sponge-like material, such as the housing polishing mat Scotch Britea. In the case of a 3,000 l tank, three pieces measuring 15 x 100 cm are hung vertically in the water (by means for example of wooden stick laid across the tank top rim), and are kept vertical by a terminal weight. The water circulation pushes particulate matter against the trap, where they adhere and clog the mat. These mats are also an ideal substrate for *Vorticella*, a sessile ciliate that competes with rotifers for food. To be effective, the traps must be cleaned at least twice a day.

To clean particle traps proceed as follows:

- Take them out of tank carefully, avoiding dripping of trapped material back to water
- Remove all particles with a high pressure jet of tap water
- Dip the trap in a 500 ppm hypochlorite solution for one hour to disinfect
- Rinse well with tap water
- Dip in a thiosulphate solution to neutralize residual chlorine & place again in the tank.

Harvest

At harvest, rotifers are filtered and rinsed before being fed to fish or utilized as inoculum for new tanks. For this purpose, a double submerged filter is used. The inner filter has a mesh size of 250 - 300 µm to retain larger particles, flocks of agglomerated food particles and ciliates which would rapidly clog the finer filter. The outer filter has a 50 µm filter mesh. Its capacity should be large enough to keep safely the whole rotifer population for the time needed to complete harvest and rinsing. Both filters are placed inside a large wheeled container full of water to avoid pressure build-up from the outgoing water that would smear rotifers against the net. A gentle air bubbling along the inner side of the filter helps to keep the filter free from clogging.

Harvesting and rinsing protocol for round-conical tanks with central drain:

1. Prepare the harvesting device, always clean and disinfected with hypochlorite
2. Inject pure oxygen into the tank to be harvested for 10 to 15 minutes to have a supersaturated medium (at 10 ppm DO) in which rotifers could safely stand filtering operation
3. Switch off aeration for 10 - 15 minutes and purge the bottom sediments
4. Into central drain fit a PVC pipe with holes at 15 cm from the drain; wait 5 minutes
5. Fix a flex hose to the bottom drainage valve and place the other end into the harvesting device
6. Open the valve and start filtering
7. Regulate the water flow so that the filter does not clog and the culture water does not overflow; do not exceed 100 l / minute
8. Stop water flow and clean the clogged pre-filter whenever necessary to avoid overflow of rotifer concentrated water
9. While harvesting, check for possible loss of rotifers through the net by sampling some filtered water with a beaker or a Petri dish
10. At the end of filtration, close the valve and rinse for 10 to 15 minutes the rotifers with filtered sterilized seawater at the same temperature as the tank of origin

Storage

Storage is necessary for maintenance of stock cultures, short term preservation of live harvested rotifers and long term preservations of dead harvested rotifers. Stocks of rotifers could be stored at 3°C for a month without special attention which enabled the recovery of cultures when necessary. Alternatively, adult rotifers can be maintained at low temperatures which discourage rapid population growth. The possibility of cryopreservation is also being studied. The utility of frozen rotifers for feeding larvae is also investigated.

High Density Rotifer Culture

Recent advances in mass culture technique yield higher rotifer densities. As an example, the procedure described below shows how to produce one billion rotifers daily with a battery of five 1 / m^3 round tanks with conical bottom.

Protocol

Inoculum: 500 rotifers / ml, harvesting at 1 000 rotifers / ml in four days. Prepare five tanks following the above mentioned procedure.

Environmental Parameters

- Oxygen: keep 7 - 9 ppm with pure oxygen integration, preferably controlled by an automatic monitoring device

- pH:7.5 - 8.5 add 10 % hydrochloric acid when needed
- TAN Total Ammonia Nitrogen:< 6 mg/l
- NH_3:< 1 mg / l
- Aeration: reasonably strong. Hang three airstones placed at 15 cm above the bottom along the tank wall for air injection, and place one in the centre at 20 cm from the cone tip for oxygen injection
- Particle traps: place four mats along the tank wall.

Cleaning Operations (Three Times a Day)

1. Remove traps, air tubing and diffusers and wash carefully with tap water
2. Stop aeration for 10 minutes and purge the tank cone through its bottom valve
3. Put back in place clean and disinfected traps, air tubing and diffusers.
4. Calculate the daily amount, distribute before the inoculum 100 g of it (100 ppm in a 1- m^3 tank), then divide the remaining quantity into three rations, using the same proportion for the whole day 0, at 4 and 10 pm and 2 am.

Feeding schedule (distributed at 2 and 8 am, 2 and 10 pm)

Steps	Age days	rot./ ml	DFR (g / million rotifers)
Inoculation	0	500	0.3430 a/
	1	750	0.2857
	2	900	0.2857
Enrichment b/	3	1200	0.2857
Harvest c/	4	1500	

a / of which 50% Protein Selco® and 50% Culture Selco®: b / Replace Culture Selco ® by Protein or DHA Protein Selco®. c / 1/3 of the harvest is used to inoculate other tanks, and 2/3 for fish larvae feeding (direct or after 12 hours-enrichment in a new tank).

Chapter 15

Biology and Culture Requirements of Artemia

Artemia is a genus of aquatic crustaceans known as brine shrimp. Artemia, the only genus in the family Artemiidae, The ability of the Artemia to produce dormant eggs, known as cysts, has led to extensive use of Artemia in aquaculture. The cysts may be stored for long periods and hatched on demand to provide a convenient form of live feed for larval fish and crustaceans.

Nauplii of the brine shrimp Artemia constitute the most widely used food item, and over 2000 tonnes of dry Artemia cysts are marketed worldwide annually. In addition, the resilience of Artemia makes them ideal animals for running biological toxicity assays and is now one of the standard organisms for testing the toxicity of chemicals. A breed of Artemia is sold as a novelty gift under the marketing name Sea-Monkeys.

Artemia strains populate all over the world, more than 300 salt lakes. Usually, due to high osmoregulation ability, the artemia live in waters with extreme salinities The artemia can resist low oxygen levels, due to capability of synthesizing respiratory mechanism (hemoglobin). Artemia adapted to wide range of temperatures (6 - 35 ^{0}C). Artemia are non selective filter feeders.

Phototactism and Distribution

Young naupliae are positively phototactic. Adults are negatively phototactic. Artemia tends to gather in clouds along the shores. Vertical and horizontal distribution patterns are completely different during day and night time.

Brine Shrimp Fact Sheet

Under optimal conditions, brine shrimp grows from larvae to adult in less than 2 weeks, increasing in length by a factor of 20 & in biomass by a factor of 500. Artemia can be found in a wide range of water salinities i.e. from 10 ppt to saturation level high fecundity rate (more than 100 cysts or nauplii, every four days) Long life span (exceeding 6 months). Above 100 ppt, no predators or food competitors survive, resulting in a monoculture under natural conditions.

Artemia Taxonomy

Artemia, a primitive brine shrimp that was found as an optimal live food for aquatic animals, which are cultured in aquaculture systems

PHYLUM	:	Arthropoda
CLASS	:	Crustacea
SUBCLASS	:	Branchiopoda
ORDER	:	Anostraca
FAMILY	:	Artemidae
GENUS	:	Artemia (Leach, 1819)
SPECIES	:	*Artemia salina*

Distinction between oviparous eggs (cysts) and ovoviviparous eggs (will develop into naupliae)

Cysts	***"nauplius"eggs***
hard and dark shell	no shell, soft
brown shell gland	whitish shell gland

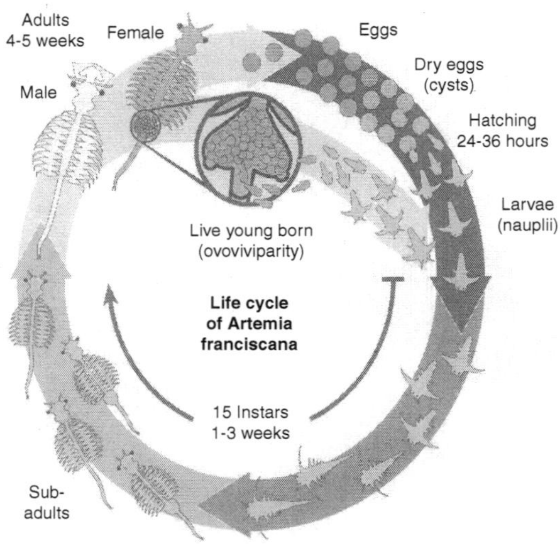

Reproduction

Adult female brine shrimp ovulate approximately every 140 hours. In favourable conditions, the female brine shrimp can produce eggs that almost immediately hatch. While in extreme conditions, such as low oxygen level or salinity above 150 ‰, female brine shrimp produce eggs with a chorion coating which has a brown colour. These eggs, also known as cysts, are metabolically inactive and can remain in total stasis for two years while in dry oxygen-free conditions, even at temperatures below freezing. This characteristic is called cryptobiosis, meaning "hidden life". While in cryptobiosis, brine shrimp eggs can survive temperatures of liquid air (−190 °C or −310 °F) and a small percentage can survive above boiling temperature (105 °C or 221 °F) for up to two hours. Once placed in briny (salt) water, the eggs hatch within a few hours. The nauplius larvae are less than 0.4 mm in length when they first hatch. Brine shrimp have a biological life cycle of one year. In normal conditions the reproduction is sexual (Ovoviviparous). In extreme conditions the reproduction is asexual resulting with dormant eggs (Oviparous).

REPRODUCTION

OVIPAROUS	OVOVIVIPAROUS
low O_2-content (such as in high salinity)	high oxygen content (such as in low salinity)
strong O_2-fluctuations	minor O_2-fluctuations
Fe-rich food (such as gree algae)	Fe-low food (such as organic debris)

Oviparous Reproduction

After copulation fertilized eggs are surrounded in the brood pouch of the female with a tough brown shell. The egg is then called a cyst. The cysts are released by the female in the water where they will not hatch until they have been completely dehydrated (in nature by floating ashore and sun-drying).

The embryo inside each cyst is then in a state of metabolic dormancy and will not further developing until hydrated again (water absorption). In nature this mostly happens when washed back to the pond water after rainfall. Once sufficiently hydrated the embryo further develops into the instar I larva (nauplius) which will hatch out of the cyst shell. In order to result free fins nauplia from cysts, they need water (hydration) and oxygen to initiate and complete the metabolism.

Ovoviviparous Reproduction

After fertilization the eggs are not surrounded by a shell but instead immediately develop further into naupliae in the brood pouch of the female. These naupliae are then released in the water as free-swimming naupliae. Eggs hatch nauplia which will have a whitish colour. Nauplia are free fins larvae. In order to result free fins nauplia from cysts, they need water (hydration) and oxygen to initiate and complete the metabolism. Cysts (0.2 - 0.3 mm) turn into larvae water swimmer, nauplia (0.45 mm) in a period of 24 - 36 hours. Complete hydration cyst takes about an hour. Depending on food availability, nauplia will reach adults (max. 12 mm) within 3 weeks. Dehydrated cysts - on average measure between 200 and 270 µm weigh 3.5 µg. Dry cysts are very resistant to extreme conditions. Up to 80°C, hatching efficiency is not affected. Hydrated cysts are killed by temperatures lower than 0°C and higher than 40°C.

Cysts

Cysts are very hygroscopic and absorb even water from the atmosphere. When stored, water content should be lower than 0.09 g H_2O / g cyst or 10 % (no metabolism). At 0.3 g H_2O / g cyst or 25 %, the metabolism of the dormant cyst starts.

At salinities higher then 70 ppt cysts cannot hatch because of the too high osmotic gradient. In salinities lower than 5 ppt cysts will hatch but resulting naupliae will die quickly. Light triggering is needed at the beginning of the hatching to start metabolism. In freshwater and seawater, dehydrated and hydrated cysts sink. In brine (saturated salt solution) they float.

Diet for Artemia

An Artemia cyst In their first stage of development, Artemia nauplii do not feed but consume their own energy reserves stored in the cyst Wild brine shrimp eat microscopic planktonic algae. Cultured brine shrimp can also be fed particulate foods including yeast, wheat flour, soybean powder or egg yolk. Artemia is an obligatory non-selective particle filter feeder and removes suspended particles smaller than 40 – 60 µm down to a few µm from the water with great effectiveness. Food particles may consist of algae cells (non-filamentous), Protozoa, organic detritus particles, etc. Only from instar II and III larval stages food is taken up. Instar I through III larvae has a yolk reserve. In most waters, where, Artemia occurs naturally, no other plankton feeder is present. Therefore, the only food competition will be among the Artemia they.

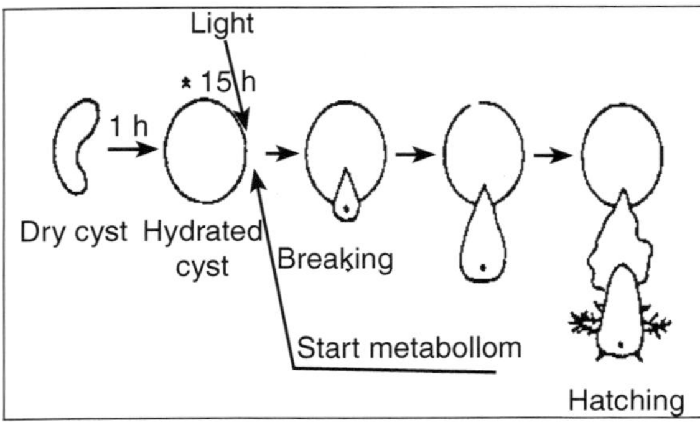

Development of an Artemia cyst from incubation in seawater until nauplius release

Advantages of Tank Production and Tank Produced Biomass

Although tank-produced Artemia biomass far more expensive than pond produced brine shrimp, its advantages for application are manifold.

Year round availability of on grown Artemia, independent of climate or season

Specific stages (juveniles, pre adult, adults) or prey with uniform size can be harvested as a function of size preferences of the predator;

Quality of Artemia can be better controlled (e.g. nutritionally, free from diseases)

The abiotic and biotic conditions relevant for Artemia culture are: physico-chemical culture conditions, ionic composition of the culture media, temperature, salinity, pH, oxygen concentration, illumination, water quality,

Culture Density of Artemia

Artemia can be cultured at high to very high densities without affecting survival. Depending on the applied culture technique, inoculation densities up to 5,000 larvae per litre for batch culture, 10,000 for closed flow through culture and 18,000 for open flow through culture can be maintained without interference on survival.

Culture Techniques- Two Types

A. Saltwater ponds

B. Controlled systems Tanks, Raceways, Flow through systems

Production in Saltwater Ponds

Large scale culture of artemia can be carried out in manmade saltwater ponds for biomasss & cysts production. Such a system is termed as extensive culture. In Brazil, China, Philippines, Thailand and Mexico, several thousands of hectares of salterns which unsuited for classical agriculture due to salinity problem, having salinity 30 to 200 ppt and water temperature 25 to 35°C used for biomass and cyst production.

The basic principal for artemia production in saltern: Inoculation of small quantity of freshly hatched nauplii (1 to 10 g / l) after deepening the ponds to 40 to 50 cm and ensuring that predators are excluded. The ponds are fertilized to augment enough food production to maintain continuous ovoviviparous mode of reproduction to facilitate fast increase in population density to harvest biomass and further exposing artemia to higher salinity level (> 150 ppt) give stress to the population thereby helps in induction of oviparity to produce cysts. The liberated cysts will float and due to wind action they will accumulate on shore and from they can be harvested. A production of 15 g live weight / m^3 / day 60 kg dry weight / ha / 5 months is usually achieved from salt pans.

Biomass production of artemia under controlled condition (intensive) can be carried out in batch (Tanks and Raceways) or in flow through culture systems with provisions for maximum oxygenation ensuring continuous food availability.

Batch Culture

Outdoor rectangular tanks with curved corners a central partitioning are used without any water renewal in air water lift raceway at 50 - 100 ppt saline water to avoid competitors and predators in the system. To this, freshly hatched nauplii are stocked at a rate of 10,000 numbers / litre feeding is done with wide range of feed. Both live inert materials (rice bran, wheat flour, yeast) of less than 50 µ. Faecal pellets and exuviae have to be regularly removed from 4th day of culture onwards pH of water should be maintained > 7.5 Water depth should not exceed 1 meter.

Generally, the culture period is for about two weeks. Harvesting is done by scoop net after stopping aeration. A biomass production of 5 kg / cubic meter in 2 weeks can be obtained. Flow through culture system. More intensive artemia culture can be achieved with flow through system in which continuous renewal of culture medium with food (micro algae or inert diet) to the culture tank is maintained.

Flow through Culture System

The continuous water exchange ensure the removal of all metabolites thereby intensive stocking i.e. double the stocking rate practiced in batch culture (20000 numbers / litre) To achieve a production rate of 25 kg / m^3 / 2 weeks. By giving salinity and oxygen stress, the production of cysts is carried out in intensive system liberated cysts are collected in 10 micron filter bags.

Rate of production - Batch culture: A production of 5 kg Artemia biomass (wet weight) per cubic metre of water during a period of 2 weeks will be obtained when stocking is done at the rate of 10,000 nauplii per litre of the medium feeding with rice bran at 15 - 20 cm transparency.

Rate of production - By flow-through culture system, a production of 25 kg biomass m^3 / 2 weeks will be achieved by stocking 20, 000 nauplii / litre feeding with rice bran or algae at 15 - 20 cm transparency.

Harvesting and Processing for Storage

Cysts from salt pans are collected with the help of double screen dip net from the corner of ponds in the morning harvested cysts are stored in saturated brine before processing, cleaning and drying. Raw cysts are then dried cured and packed under vacuum / nitrogen atmosphere. Adult live artemia can be manually harvested with a dip or conical bag net with less than 100 μ mesh. Always frozen in live condition to maintain nutritional quality.

Control of Infections in Artemia Culture

Occasionally heavy losses of pre adults are faced due to infections with filamentous Leucothrix bacteria which occur especially in nutrient-rich media. The Leucothrix colonies fix on the exoskeleton by preference on the thoracopods, become only visible from the instar V/VI-stage onwards. Eventually, growth and molting are arrested, overfeeding of the tanks occurs, resulting in collapse of the Artemia culture. Application of terramycin is a method used to control this. The most practical solution appears to be a raise in salinity from 35 to 50 - 60 ppt together with a higher water renewal rate.

A second observed disease in Artemia cultures so called 'black disease where part of the animals show black spots especially at their extremities, i.e. thoracopods, antennae as a result detachment of the epidermis from the cuticula occurs, caused by a dietary deficiency which interferes with lipid metabolism. In high density culturing of Artemia using agricultural by-products as a food source, the black disease is observed when water quality deteriorates or when feeding rates are not optimal. Suitable conditions for hatching Artemia are temperature (25 - 28 °C), salinity (5 - 35 ppt), Heavy continuous aeration, Constant llumination (two 40 watt fluorescent bulb), and pH (7.5 - 8.0). Stocking density is set by adding not more than 5 grams of cysts per liter of water. Good circulation is needed to keep the cysts in suspension. A container that is V - shaped or Cone - shaped is best. Unhatched cysts empty shells hatched nauplii can be easily removed separately from the cone shaped bottles or tanks. Usually, the hatching percentage and density is a function of water quality, circulation, the origin of the cysts.

Requirements and Preparation of Artemia

The important step to be followed for preparation of Artemia for feeding larval rearing in the hatchery:

Determine weight of Artemia cysts required to feed larvae in a tank of known volume

Hydration and decapsulation of cysts

Incubation of cysts

Separation of cysts from shells and debris

Count the hatched Artemia

Calculate the number of Artemia remaining in the rearing tank from previous feeding

Calculate number of Artemia nauplii required by larvae & transfer them to rearing tank

Determine the amount of Artemia cysts required

To estimate the amount of Artemia required for the larval rearing, one must consider both the volume of the tank and the expected number of Artemia the larvae will consume. Artemia nauplii are maintained in the larval culture tank at densities of 2 to 5 numbers / ml for most species of finfish and up to 6 per ml in the more advanced larval shrimp stages.

Determine the Amount of Artemia Cysts Required

Each gram of cysts contains approximately 2,00,000 to 3,00,000 cysts and generally artemia have at least 50 percent hatchability. Based on this, the requirements of artemia cysts need to be hatched daily can be calculated as follows:

Weight of Artemia cysts required

$$\frac{\text{Total volume of all rearing tanks (in ml)} \times \text{No. of Artemia required per ml}}{\text{Percentage hatch} \times \text{No. of cysts per gram}}$$

Advantages of De-capsulating

Cysts must be disinfected with chlorine when using Artemia for saltwater to kill bacteria such as *Vibrio*. Cysts do not need to be hydrated before hatching, saving time. With no shells, nauplii can hatch quicker with less expenditure of energy (fat content), increasing their nutritional value. Hatching time is reduced from the normal 24 hours to 16 hours at 30°C (86°F). Nauplii lose their nutritional value over time, so the extra 8 hours saved means nauplii with higher fat content.

Quicker hatching time results in smaller, more uniform-sized nauplii suitable for the smallest sized larvae. There are no shells or un hatched cysts to separate from the nauplii. Even unhatched cysts can be eaten by larvae. There are no shells or un hatched cysts accumulating on the surface of larval tanks. Shells or unhatched cysts can lodge in the esophagus and gut and kill larvae. Hatching percentage increases compared to conventional hatching of under-capsulated cysts. It is cost efficient as the price of Artemia has doubled. As Artemia supply can fluctuate, even poorer grades (quality) of cysts (such as Grade A or B) can be used.

Disadvantages of De-capsulating

De-capsulation requires skilled labor as the window of de-capsulation is only 2 to 4 minutes. Any errors in following procedures result in loss of time and cost of Artemia cysts and chemicals. For large scale hatcheries when several cans must be hatched every day, following the disinfection / hydration techniques is more time and cost efficient.

Decapsulation of hydrated cyst needs to be decapsulated immediately using decapsulation solution. Transfer the hydrated cysts to the hypochlorite solution either liquid bleach (NaOCl - 5 ml / g of cysts) or bleaching powder CaOCl) (0.5 g active hypochlorite product per g of cysts (pH > 10 adjusted with 150 mg of NaOH / g of cysts) or For 250 mg cysts, decapsulation solution can be made with 40 ml fresh NaOCl + 20 ml seawater (5 ppt salinity) + 1.5 ml of 40 % NaOH solution added though the side of beaker. With the addition of the bleach, the reaction (decapsulation) begins. Ice can then be placed below in the solution to keep the temperature below 30°C and stirrer as fast as possible. A white foam layer will develop and the solution will change from brown to gray or light orange approximately 6 to 8 minutes.

When no more colour change is seen, the decapsulation process is completed and drains through a 100 to 125 micrometer sieve, and washes the cysts thoroughly with tap water or seawater. The washing should continue for about 10 minutes until no chlorine smell can be detected and dip the cysts into 0.1 N HCl (not more than 30 seconds) to bring the pH to the neutral range. Pour the cysts into the sieve again and wash with water for 3 minutes. The cysts are now ready for incubation. Approximately 275,000 cysts / g produce 200,000 - 250,000 nauplii / g or 90 - 113 million / 1 pound can (454 g)

1. Pre-cool (5°C) 1 gallon of salt water and 1 gallon of liquid chlorine (10%) in a refrigerator for at least 2 days.
2. Dissolve 20 grams (1/16 cup) of sodium hydroxide $Na(OH)_2$ beads in 30 ml (1/8 cup) of freshwater. Caution mixture is very hot and can burn skin. Do not inhale fumes !
3. Dissolve 1/4 cup of sodium thiosulfate in a 1 liter beaker of freshwater
4. Prepare 4 gallons of brine solution (5 lbs. of crystal salt dissolved in 5 gallon capacity bucket of freshwater)
5. 1 pound can of Brine Shrimp cysts

6. Aerate cysts for 1 hour in 15 liters of freshwater in a brine shrimp net suspended in a hatching cone
7. Rinse cysts briefly in freshwater
8. Add the $Na(OH)_2$ solution to 2 liters of chilled salt water
9. Discard remaining salt water
10. Add cysts to empty clean cone
11. Add the 2 liters of salt water mixture
12. Aerate vigorously
13. Add 2 liters of chilled bleach to cone
14. After 2 - 4 minutes, cysts will turn bright orange
15. Add 1 liter of sodium thiosulfate solution and 15 seconds
16. Turn of the air, drain, rinse with water in a brine shrimp net
17. Add 2 liters of brine solution to a clean cone
18. Add cysts and rinse net with more brine and fill up to the line for the required number of cups. For example, the number of cups (18) times 200 ml = 3.6 liters water and cysts
19. Aerate slowly for 18 hours
20. With continued aeration, drain cysts into 16 to 24 (average 18) - 200 ml containers
21. Refrigerate for up to 2 weeks

Hatching Procedure

The following steps will achieve optimum brine shrimp hatch rates:

Set Up: Place hatching cone or similarly shaped vessel. Cone should be semi-translucent for ease of harvesting and light transmission

Add Water: Fill cone with water and adjust salinity to 25 ppt Optimum hatching temperature is 82°F (28°C)

Add Cysts: Add cysts at the rate of 1 gram per liter

Aerate: Provide adequate aeration to keep cysts in suspension

Hatch: Depending upon water temperature, cysts should hatch in approximately 18 - 36 hrs

Harvest: After hatching brine shrimp, turn off or remove aeration and wait several minutes for the shells and baby brine shrimp (or nauplii) to separate. Newly hatched nauplii will settle to the bottom of the cone or move towards a light source; the shells will float to the surface. Once separated, the nauplii can be siphoned from the bottom with a length of air tubing or gently drained through the bottom of the cone through a valve, if so equipped.

Rinse: The warm incubation temperatures and metabolites from the hatching medium create ideal conditions for a bacteria bloom. Rinsing of the baby brine shrimp in a fine mesh net or sieve using clean fresh or salt water is important before feeding them to your fish.

Clean Equipment: Tanks and brine shrimp hatching equipment should be cleaned and disinfected routinely.

Harvesting: Separation of Cysts from Shells

After 15 to 20 hours of incubation, most of the cysts will be hatched and there will be a noticeable colour change in the culture from brown to orange. At this time, stop the aeration and the pinkish orange nauplii will be seen swimming in "clouds." **Harvesting:** Separation of cysts from shells any empty or undissolved shells tend to float, while the unhatched cysts and debris sink. If the container is clear, a black cloth can be placed on top of the container to speed up the concentration of nauplii toward the bottom.

Harvesting: Separation of Cysts from Shells

A siphon also can be used to remove first the debris and then the nauplii from the bottom. The nauplii should be collected on a 100 to 120 micrometer screen, washed with clean seawater and placed in a small volume of water. Washing removes contaminants and hatching metabolites. The phototactic method also can be used to harvest the newly hatched nauplii. Wash the harvested nauplii for feeding.

Counting of Hatched Artemia

The contents of this beaker should be increased to 100 ml with distilled water (1:100 dilutions). Add full strength Lugol's stain or iodine to the beaker to slow down or kill the nauplii if high concentrations are present. Mix the contents of the beaker well and draw a 1 ml sample with a pipette for counting. If the numbers are high, count the nauplii in 0.1 ml. This count gives the concentration of nauplii in 0.1 ml of the dilution (H). Multiply the number counted by ten to get the number of Artemia in 1 ml and then multiply by the dilution factor (100) to calculate the number of Artemia / ml. To get the total number of Artemia using the formula:

Number of Artemia / ml concentrated in 00 ml beaker (G)

H x 1000 (where H = number in 0.1 ml)

Calculation of the Number of Artemia in the Rearing Tank

Scoop a 100 ml of sample from the rearing tank into a beaker and add Lugol's stain or iodine to kill the Artemia. Use above procedure, but do not dilute the sample.

Calculation of the Number of Artemia in the Rearing Tank

The concentration of artemia in the rearing tank can be calculated using the following formula

Concentration of Artemia per ml (I) = number per 0.1 ml (J) x 10

Calculate the Number of Artemia to be Fed

The quantity of Artemia added to the larval system must be adjusted on the basis of the number remaining from the previous feeding. Each day, the stage of the fish larvae is assessed and used to determine the total quantity of Artemia required per ml of the rearing tank (K). From this subtract the number of Artemia per ml present in the tank (H) from the total you need to find the quantity / ml to be added. The quantity to be added (L) equals (K) minus (J). Multiply this number by the volume of the tank in ml to find the total number you require in the tank (M). Divide this number by the number of nauplii per ml in the fresh batch of Artemia (G) to determine the number of ml to be transferred to the fish tank. It may be easier to calculate with the following formula:

$$V = \frac{(K - H \times 1000) \times M}{10 \times J}$$

Cyst Quality Analysis

Cyst quality can be assessed on the basis of its moisture content hatching efficiency hatching percentage hatching rate besides its nutritional value. The quality of Artemia cysts mainly depends on the way the cysts are processed and stored. During prolonged period of cyst storage, periodical analysis is necessary to make sure that the cysts are retaining their viability Hatching efficiency (HE): Number of Arfemia nauplii hatched out of 1 g of cysts under normal hatching conditions (i.e., 48 hours incubation in 35 ppt seawater with 8.0 - 8.5 pH at 25 - 29°C). It can be calculated by applying the following formula.

HE = N x 4 x I00 x 4 where N is the average number of Artemia nauplii present in 0.25 ml sample.

Procedure to Find out Hatching Efficiency (HE)

Take 80 ml of sea water in 100 ml measuring cylinder. Aerate and add 250 mg Artemia cysts to be anaiysed. After one hour, adjust to 100 ml by adding sea water. Provide continuous aeration and illumination. After 48 hours, take 5 subsamples of 0.25 mi each in petri dishes with 1 rnl graduated pipette. Add 1 drop of Lugol's solution. Count the nauplii per petri dish and make an average (N).

$$\frac{\text{Number of naupli}}{\text{One of cyst}} HE = N \times 4 \times 100 \times 4$$

Hatching Percentage (H %)

It reveals the number of nauplii obtained out of 200 full (embryo containing) cysts. Treatment with sodium hypochlorite after 48 hours dissolves the shell thereby exposing the unhatched embryos from the unhatched cysts.

Procedure to Find out Hatching Percentage (H %)

Take 80 ml of sea water in 100 ml measuring cylinder

Aerate and add 250 mg Artemia cysts, to be analysed

After one hour, adjust to 100 ml by adding sea water. Provide continuous aeration and illumination

After 48 hours, take 5 subsamples of 0.25 ml each in petridishes with 1 ml graduated pipette

Add few drops of sodium hypochlorite solution

Count the nauplii and unhatched embryos in each petridish and make an average. Hatching percentage can be calculated by making use of the following formula:

$$\text{Ho/o} = \frac{\text{Number of naupli}}{\text{Number of naupli + Number of unhatched embryos}} \times 100$$

Properly processed cysts will have more than 90 % hatching as hatching percentage depends mainly on the nature of processing and storage.

Hatching Rate (HR)

4 Indicators for Determination of Brine Shrimp (Artemia) Egg Hatch Rate Quality

This is ascertained by first determining the Cyst Per Gram (CPG) count and then determining the Nauplii Per Gram count and dividing the two as NPG / CPG.

This number is important because it is the industry standard by which brine shrimp eggs are sold. However, as we are dealing with a living organism, there are inherent variations with the physical characteristics of brine shrimp eggs. Variations that may affect the HR % can be attributed to many factors including the geographical source of the eggs, the biological strain of the eggs, processing methods, etc.

Nauplii Per Gram (NPG)

This is determined by obtaining the total number of newly hatched nauplii within the hatching container and dividing this number by the number of grams of brine shrimp eggs that were initially put into the hatching container.

This number is significant because this is the number of brine shrimp nauplii individuals that the customer is getting for their money. It is also significant because this is the end product, i.e., this is what will actually be fed to, or consumed by, the fish or shrimp larvae. Many hatchery technicians base their feeding regimes on this number.

Separation

This is a subjective indicator ascertained by the individual or technician in charge of hatching out the brine shrimp eggs. It is determined by first taking a sample of approximately 500 - 1000 ml in an imhoff cone or glass beaker from the hatching container immediately prior to harvesting the brine shrimp nauplii. After the sample has been taken, a visual observation is noted as to the separation of newly hatched brine shrimp nauplii from un-hatched eggs and shells. With excellent separation, one should observe mostly newly hatched brine shrimp nauplii, with little to none eggs or shells on the bottom of the cone or beaker.

This indicator is important because the introduction of un-hatched eggs and shells into the larval rearing tanks can cause water quality problems leading to health related problems with the cultured organisms.

Volume (CC's)

This is an indicator that determines the "volume" of a given amount of newly hatched brine shrimp nauplii within a specified period of time. SCI ascertains this in the following way: Brine shrimp eggs are allowed to hatch in an Imhoff cone and after 24 hours, the aeration is removed. After two minutes without aeration, the amount of brine shrimp nauplii are observed and measured at the bottom of this Imhoff cone as per the graduations (in cc's) located on the Imhoff cone itself. This indicator is important to those individuals that determine hatch rate quality visually with emphasis being placed on how much "volume" the brine shrimp nauplii are occupying within the container they are using to observe the newly hatched nauplii. Again, the biological variations associated with this living organism are what cause this indicator to vary.

Hatching Rate (HR)

It refers to the time period from the start of incubation (hydration of the cysts), till the completion of the release of all nauplii (completion of hatching). To find out the hatching rate (HR), hatching efficiency (HE) is to be calculated in the interval of three hours during the whole hatching period from the initial 9 hours. From this, TO, Tlo and TgO for the cysts, to be analyzed, can be worked out. It can be observed that they vary from strain to strain and even batch to batch of the same strain. The following time intervals can be worked out. To is the incubation time till the appearance of the first free swimming nauplii. Tjo is the incubation time, taken for the appearance of 10% of the total hatchable nauplii. TgO is considered as the incubation time, needed for the appearance of 90% of the total hatchable nauplii.

Procedure to Find out Hatching Rate (HR)

Take 80 ml of sea water in 100 ml measuring cylinder

Aerate and add 250 mg Artemia cysts, to be analyzed

After one hour, adjust to 100 mi by adding sea water

Provide continuous aeration and illumination

Starting from 9 hours incubation, take 5 subsamples of 0.25 ml each, in five petri dishes and add I drop of Lugol's solution in each petridish

Repeat the sampling at every three hours i.e., 12, 15, 18, 21, 24, 27, 30, 33, 36, 39, 42, 45 and 48 hours incubation (totally 70 petri dishes)

Count the nauplii and make an average for every 3 hours incubation after initial 9 hours

Plot the average nauplii number and time in hours as hatching rate curve From the curve, To, Tlo, TsO and TgO can be extrapolated

Chapter 16

Spirulina Cultivation

Application

Food Supplement

The World Health Organization has found Spirulina to be an excellent food for human consumption. Spirulina has the approval of the Food & Drugs Authority of the United States for being sold as a natural food. In Japan and in the United States, business executives take spirulina tablets to combat stress. Athletes and joggers take spirulina for quick energy synthesis.

Health And Medicine

Non insulin dependent diabetes, Cholesterol control, Vitamin A, deficiency & malnutrition, Adjunct to cancer patients undergoing chemotherapy, Formulations with other natural products as a general health supplement., Liver corrective for liver disorder, Burns therapy, skin grafting, Control obesity, Lactating agent for mothers.

- Reduces blood cholestrol
- Spirulina helps to control diabetes
- Helps in kidney detoxification
- Useful in combating malnutrition
- Spirulina is very useful in preventing vitamin A deficiency (helps prevent blindness and eye disease)
- Aids in the prevention of anaemia

Application - Extraction

Beta-carotene for medicinal & laboratory use, C- phycocyanin colouring agent in food, microbiological areas cosmetics, C- phycocyanin - colouring agent in food, cosmetics, etc. Chlorophyll - colouring agent, Essential aminoacids - for microbiological & chemical essays.

Application - In Pisiculture

Speciality feed for aquarium fish, Colour enhancement feed for Gold fish, Formulation with existing feeds for augmentation of vitamins, High protein feed for table variety fishes (fresh water), and Special feed for shrimp farming.

Spirulina As A Nutritional Supplement

Spirulina can be used as a partial supplementation or complete replacement for protein in aqua feeds. It was found that the growth of silver sea bream (*Rhabdosargus sarba*) fed on spirulina meal at up to 50 % level gave good results At 75 % inclusion level, growth reduces significantly, but feed conversion efficiency still comparable to the control diet, which sharply reduces at 100 % inclusion level. It is better to replace 50 % fishmeal by spirulina meal so that there will be no adverse effect on growth.

Application In Entomology And Sericulture

Feed to increase yield in mulberry consuming silk worm, Speciality feed for breeding and culturing various insects used for research.

Application In Cosmetics

Spirulina in pimple lotions, Facial asks, Hair oil, Shampoo, Mineral bath, Skin cleaner, Tooth paste.

What is called "spirulina" Here actually bears the scientific name of *Arthrospira platensis Arthrospira maxima* a cyanobacteria. But, the common name "Spirulina" is universally used.

Morphology

Spirulina is symbiotic, multicellular and filamentous BG microalgae with symbiotic bacteria that fix N_2 from air. Spirulina can be rod - or disk - shaped their main photosynthetic pigment "phycocyanin" which is blue in colour. These bacteria also contain chlorophyll a and carotenoids. Some contain the pigment "phycoythrin" giving the bacteria a red or pink colour. Spirulina are photosynthetic and therefore "autotrophic" Spirulina reproduce by binary fission. The presence of gas-filled vacuoles in the cells, together with the helical shape of the filaments, result in floating mats. Body surface of spirulina - Smooth and without covering so it easily digestible by simple enzymatic systems.

The basic biochemical composition of Spirulina Protein: Spirulina contains unusually high amounts of protein, between 55 and 70 % by dry weight, It is a complete protein, containing all essential amino acids, though with reduced amounts of methionine, cystine and lysine, as compared to standard proteins such as that from meat, eggs, or milk; it is, however, superior to all standard plant protein, such as that from legumes.

Essential fatty acids: Spirulina has a high amount of PUFAs, 1.5 - 2.0 % of 5 – 6 % total lipid. In particular, spirulina is rich in -linolenic acid (36 % of total PUFAs) also provides -linolenic acid (ALA) linoleic acid (LA, 36 percent of total) stearidonic acid (SDA) eicosapentaenoic acid EPA) docosahexaenoic acid (DHA) arachidonic acid (AA)

Vitamins: B1 (Thiamine) / B2 (Riboflavin) / B3 (Nicotinamide) / B6 (Pyridoxine) B9 (Folic acid) / B12 (Cyanocobalamin) / Vitamin C / Vitamin D / Vitamin E.

Minerals: Rich source of potassium and also contain calcium, chromium, copper, iron, magnesium, manganese, phosphorus, selenium, sodium, zinc, allophycocyanin.

Photosynthetic pigments: contains many pigments chlorophyll a xanthophyll betacarotene echinenone, myxoxanthophyll zeaxanthin canthaxanthin diatoxanthin 3 - hydroxyechinenone beta - cryptoxanthin oscillaxanthin Plusphycobiliproteins c - phycocyanin & allophycocyanin.

Nutritional composition of Spirulina -Content of its dry weight

Macro Nutrients	*Important Micronutrients*
Protein: 60 - 70 %	Minerals: 6 - 9 %
Carbohydrate: 15 - 20 %	Phycocyanin: 140 - 200 mg/ g
Fats (Lipids): 5 - 6 %	Beta carotene: 2.1 - 2.4 mg /g
Moisture: 3 - 5 %	Iron: 1.0 - 1.2 mg /g

Natural Production

Most commercial production systems are based on shallow raceways in which spirulina cultures are mixed by a paddle wheel. However, there are still some examples of spirulina being harvested commercially from naturally occurring populations (Mexico).The so-called "semi-natural" cultivation process consists of harvesting during day and night and the algal biomass doubles in three to four days. After filtration, the algal biomass is spray-dried afterhomogenization and pasteurization, the first pilot plant which produced 150 tonnes of dry spirulina biomass per year started production in 1973; its production capacity was thereafter raised to 300 tonnes of medium-grade product / year from 12 hectares of natural ponds. Another semi-natural lake in Myanmar has been reported to be used as a production site for spirulina. Four volcanic lakes with natural spirulina blooms were studied beginning in 1984. Production began at Twin Taung Lake in 1988, and by 1999 increased to 100 tonnes / year. About 60 % is harvested from boats on the surface of the lake and about 40 % is grown in outdoor ponds alongside the lake.

Laboratory Cultivation

Eight major environmental factors influence the productivity of spirulina: Luminosity (photo-period 12 / 12, 4 luxes), temperature (30 °C) inoculation size stirring speed dissolved solids (10 - 60 g / litre) pH (8.5 - 10.5) water quality macro and micronutrient presence (C, N, P, K, S, Mg, Na, Cl, Ca / Fe, Zn, Cu, Ni, Co, Se)

Spirulina cultivation - number of advantages over traditional agriculture:

High yield: With around 60 % protein content, spirulina's rapid growth means it yields 20 times more protein per unit area than soybeans40 times more than cornover 200 times more than beef

Soil requirements: Spirulina culture does not require fertile land and can actually benefit from saline conditions.

Efficient water use: Spirulina uses less water / kg of protein (approx. 2 100 l / kg protein) than other crops. Water can be recycled and the only significant water loss is through evaporation.

Spirulina culture uses 25 % of the water of soy, 17 % of corn and 2 % the water required for beef protein. Brackish or saline water can be utilized.

Efficient energy use: Spirulina requires less energy input / Kg than soy, corn or beef, including solar & generated energy. Its energy efficiency (food energy output / kg / energy input / kg) is 5 times higher than soy, 2 times higher than corn,

In India, Murugappa Chettiar Research Centre in Chennai Developed the technology successfully propagated on a large scale in the rural areas of Pudukottai district of Tamil Nadu. Domestic cultivation method "Mud Pot Spirulina Cultivation"

The method of working is very simple. This method requires only 3 mud pots each pot - 35 to 40 liters capacity an exposed but protected open area. For instance, "mud pot" spirulina production uses a medium consisting of biogas slurry (very cheap and easily available) 2 - 3 g of sea salt or chemical medium (potassium dihydrogen phosphate, cooking soda and sodium chloride). Pure spirulina culture. Three pots are buried till the neck in the ground. Filled with water mixed with the medium. A small quantity of pure spirulina added kept in sunlight and need to be stirred at least 4 times a day as the spirulina cannot grow in stagnant conditions. After 3 - 4 days of maturation the Spirulina is ready. Mature spirulina when pale medium turns intodark green

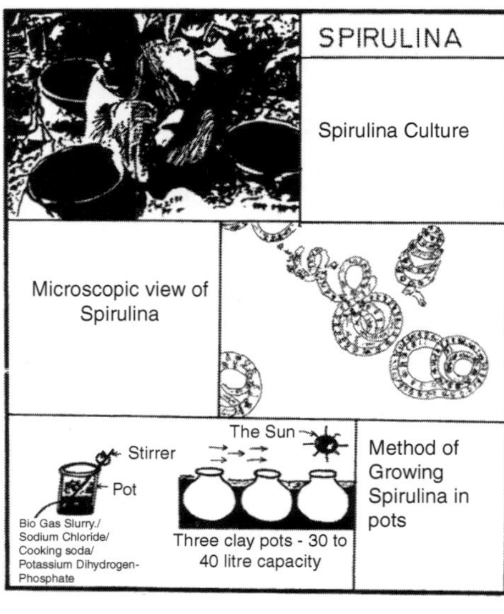

Harvested by a simple clean cloth filtration. Washed in fresh water. Spirulina immediately used for consumption or if a powdered form is desired it should be dried immediately. After washing the spirulina in fresh water (to remove adhering chemicals), it can be directly mixed with chapatti dough, chutneys, noodles, pulses, vegetables etc. Spirulina can also be preserved by drying, which must be conducted immediately in order to preserve its quality and value.

Mass Cultivation of Spirulina

Shallow ponds equipped with paddle wheels to mix the culture

Two types of open raceway ponds used.

First Lined by concrete & expensive one

Second Shallow earthen tunnel lined with polyvinyl-chloride (PVC) or some other durable plastic material.

Increases the cost of production search for cheaper material and processes, such as low - cost clay sealing

Surface area of commercial raceways 0.1 to 0.5 hectares / Depth (15 to 18 cm)

The paddle wheel - most common stirring device large paddle wheel (dia up to 2.0 m & a speed of 10 rpm) or Small paddle wheel (dia of 0.7 m and a speed 2 to 3 times faster than 2.0 m diameter paddle wheel)

Harvesting, Processing and Packing - 8 Principle Stages

Filtration and Cleaning

A nylon filter at the entrance of the water pond is needed basically two types of screens Inclining or Vibrating Inclined screens 380 - 500 mesh with a filtration area of 2 - 4 m^2 / unit capable of harvesting 10 - 18 m^3 of spirulina culture / hour. Biomass removal efficiency is high (up to 95 %)two consecutive units are used for harvesting up to 20 m^3 / hour from which slurry (8 - 10 % of biomass) is produced.

Pre-concentration - To obtain algal biomass which is washed to reduce salts content

Concentration - To remove the highest possible amount of interstitial water (located among the filaments)

Neutralization: To neutralize the biomass with the addition of acid solution

Disintegration: To break down trichomes by a grinder

Dehydration by Spray-drying: This operation has great economic importance since it involves about 20 - 30 % of the production cost

Packing: It is usually in sealed plastic bags to avoid hygroscopic action on dry spirulina

Storage: Stored in fresh, dry, unlit, Pest free and hygienic store rooms to prevent spirulina pigments from deteriorating

Details for Setting Up a Spirulina Unit

Multiplication and utilization of spirulina is easy requiring less energy inputs and infrastructure. This makes it a suitable small scale enterprise that can be easily picked up by rural educated entrepreneurs for generating additional income. The activity can also be popularized among self-help groups or as cooperatives.

Raw Material: Selected strains (algal culture) and nutrients, Sodium bi-Carbonate, Magnesium Sulphate, Super phosphate, Sodium Chloride, N, P, K and Trace elements. Nutrient Mix: Each algae sp. has its own specific needs, which are carefully standardized.

Plant and Machinery: Raceways ponds, agitation system, Carbon dioxide supply system, Nutrient doser, harvester, spray dryer, packaging system, colloid mill and walk in cooler.

Other equipment: Storage bins with lid, Trolleys, Balance, handling vessels for raw material handling.

Project Cost (Base - Estimate for a Model Project)

1. Land & Land development (13500 m^2) 9,38,000.00
2. Building and civil works (100 m^2) 60,04,000.00
3. Miscellaneous fixed assets 1,01,98,000.00
4. Plant and machinery 7,70,000.00
5. Preoperative expenses 17,75,000.00
6. Total Fixed Capital 1,96,85,000.00
7. Working capital margin 4,98,000.00
8. Total Project cost 2,01,83,000.00
9. Total working capital required at 10 % turnover 11,25,000.00
10. Promoter's contribution 67,50,000.00
11. Term loan 1,34,33,000.00

Production Capacity

Economic Capacity: 50 Kg / day

Working capacity: 3 shifts / day

(300 working days / annum)

Installed capacity: 15 t / annum

Optimum capacity utilization: 70%

Chapter 17

Earth Worm Cultivation

Raising earthworms is one of the easiest ways you can go green and reduce the amount of trash entering our landfills and water systems. Earthworms can produce more compost, in a shorter time, with less effort, than any other tool known to the gardener or farmer.

Culture of Earthworms for Bait or Fish Food

These earthworms are also used for composting sewage sludge and manure, and as a dietary supplement or ornamental fish or other difficult-to-raise fish species. Earthworms are excellent food for cultured fish species.

The Structure of Earthworms

Earthworms are externally segmented (with corresponding internal segments) with no skeleton. They have a thin coloured cuticle bearing setae (bristles). They are designed as a "tube within a tube" - the digestive tract runs the length of the body from mouth to anus, with the organs of reproduction, water balance control, and nervous system control lying between the 2 tubes. Strong sets of muscles make up much of each of the inner and outer "tubes". A blood-like fluid fills the rest of the space between the two tubes

The Physiology of Earthworms

Different species of earthworms are adapted to different climates (temperature and moisture) and to different soil types (high organic carbon content) to mineral soils (very low carbon content). These adaptations are responsible for a wide range of physiological differences between earthworm species. However, there are some general comments to be made:

1. Earthworms respire ("breathe") by absorbing oxygen through their body surface into the body and sending carbon dioxide out. The oxygen dissolves in the mucous on the body surface and then passes into the highly

branched capillary blood vessels in the body wall where it is picked up by the respiratory pigment and transported to the internal organs.

2. Earthworms may survive for considerable lengths of time in water if the dissolved oxygen level is high enough.
3. Earthworms cannot maintain a steady body temperature like humans or dogs or cattle. This means their respiration rate rises with increasing temperature. Increasing respiration rates mean increased metabolic rates which means increased energy requirements which means increased feeding rates by worms or food must be "stolen" from reserves in the worms' own body tissues. The reverse is true for decreasing temperatures.
4. Earthworms derive their nutrition from organic matter in a wide variety of forms. So far plant matter (various forms, fresh-decayed), protozoans, rotifers, nematodes, bacteria, fungi and decomposing remains of other animals are known to be worm food.
5. For the manure worm, Eisenia foetida, there is strong evidence that protozoans form the basis of their diet. If this is true, it means successful culture requires that protozoan populations be high in the "food medium" whether that "food medium" be shredded paper, manure or waste foods.
6. Excretion of metabolic products (wastes of digestion, growth and death of cells) is accomplished through the nephridia (similar function to kidneys) and through the gut.
7. Blood circulates by flowing forward along the dorsal collecting vessel into pulsating hearts where it is pumped down to the ventral vessels where it is pumped forward to the head and backward to all parts of the body.
8. There is a large ventral nerve cord with an anterior enlargement in the head ("brain"). The nerve cord connects various sense organs in the body and controls muscle contraction and relaxation. Light sensitive structures have been identified in worm species. Earthworms are also very obviously capable of sensing differences in acidity, relative humidity, touch, foods (chemicals).
9. Same species are able to regenerate amputated segments of the body. Regeneration of tail segments occurs more readily than head segments. There does seem to be a limit to the number of segments that can be regenerated, but this number varies from species to species. No known worm species, if cut in half, will form 2 viable worms.

The Reproduction of Earthworms

Earthworms may reproduce biparentally (by exchanging genetic material with another worm of the same species only) or uniparentally (no sexual fertilization by another worm takes place). The method of reproduction is characteristic of the species (that is both methods of reproduction are not normally found in the same species - but relatively little is known about this). Where biparental reproduction occurs (dew worm and manure worm), both male and female organs occur in the same animal and are cross fertilized by the other worm (of the same species only) simultaneously. Both worms will then produce cocoons (capsules). In uniparental worms, some internal mechanism triggers production of an ovum, which is then released as a cocoon, which develops into a mature worm.

Normally each cocoon produces 1 or 2 worms (but as many as 11 in the manure worm). Depending on the species, it takes from 3 weeks to a year to reach reproductive maturity.

A word about "hybrid" worms. It is not technically feasible nor has it been demonstrated that it is possible to hybridize two different species of worms (such as the dew worm and the manure worm) to produce a "hybrid" worm. Any worm claimed to be a 'hybrid" which has been analyzed by an expert, has been proved to be a particular species of (not a "hybrid"). Three worm species are often touted as being "hybrids" - they are the manure worm (*E. foetida*), the African night crawler (*E. eugeniae*) and the grey night crawler (*A. hawayanis*). If you hear people talking about "hybrids", be immediately suspicious.

A. The Predators

Many species of birds (gulls following ploughs in field), moles, hedgehogs, foxes, toads and snakes are known to eat earthworms. Beetles, leeches, slugs and flatworms also feed on worms. Most of these are unlikely to be problem in earthworm cultures.

B. The Parasites

Bacteria, protozoa (single-celled animals), flatworms, nematodes (roundworms) and dipterous larvae are internal parasites of earthworms

Starting, Feeding and Care of Worm Cultures

Unoxidized peat is the preferred bedding and can be obtained from peat mining companies. The bed material should be loose, moist, and well aerated. To start a new worm bed, peat is placed to a depth of about 3 inches and leveled with a rake. Hardwood sawdust, ground peanut hulls, sand (< 2 % by weight) or other organic material can be mixed into the peat to minimize compaction. Small quantities of agricultural grade dolomite limestone are added gradually over a period of days to adjust the pH to about 6.5 - 6.8. Uncontaminated groundwater (total hardness <250 mg / l $CaCO_3$) is lightly sprinkled onthe bedding for a few seconds until the surface first glistens. Adult worms (about 2 pounds / 100 square feet of bed) are placed along the central midline of the bed. Within a few minutes, the worms work their way into the bedding, and in a few days they gradually spread evenly throughout the bed. Most cultured earthworms live in the top 2 inches of bedding. They are active at night and will crawl out of the beds in dimly lit or dark rooms. The worms are light sensitive (photophobic) and thus overhead illumination discourages their movements; the lights also help heat the beds in winter. A small fluorescent light about 2 feet above the bed keeps the worms in the bed without the expense of a large bank of electrical lights. The moisture content of the bed is a critical factor affecting worm production. Earthworms prefer moist but not saturated conditions. Excess water causes the food to rot and fungus to develop. This can reduce production and cause mortality if the condition persists. Excess moisture can also cause the bedding materials to cling to the worm's body, affecting respiration and production. Watering procedures vary with each site. Large commercial beds are usually lightly watered daily with a hose nozzle, preferably in the morning. Small plastic pools that have no drainage are not watered on weekends to allow the bed to dry out. This routine does not cause any perceptible change in worm production and reduces labor. Immediately after watering, a high fiber content food is broadcast by hand or mechanical applicator evenly over the bed's surface at a daily rate of approximately 0.5 pounds per square yard. Applied properly, the bed material should be faintly visible through the layer of food. The food will absorb some water and become moist, but not soggy. The worms migrate to the surface and begin to feed within the hour. Uneaten food may indicate too high a feeding rate or detrimental environmental conditions in the bed (e.g., too much or too little moisture, lack of oxygen, improper temperature). Uneaten lumps of food become moldy and attract unwanted pests; therefore, before the next feeding, any food that is uneaten should be gently worked into the surface of the bedding by hand with a tine hoe. This action also helps to aerate the bedding material, preventing anaerobic conditions. To optimize worm production and facilitate harvesting, the peat bedding should be loosened at least every month with a tine hoe. Gently turning the bed does not harm the wormsand keeps the bed oxygenated. The buildup of ammonia from wastes can retard worm production; therefore, a new layer (2-3 inches) of clean bedding is spread over the surface about every 3 months. After the new bedding is added, the worms migrate up into it. When the frame or container is eventually filled with spent bedding and becomes anoxic, the beds are harvested, and breeder worms are transferred to a new bed. Worms on the bed's surface while the bed is lit is a sign of excessive water or overcrowding. Listless and stunted worms indicate disease, fungus, or overcrowding. The worms are in good condition if, when the surface of the bedding is touched, the surface layer ripples as the worms quickly retract.

What Kind of Earthworms do I Need?

There are three types of composting worms and one type of garden worm available.

The Ridge Wiggeler

The Red Wiggler is also known as *Eisenia foetida*, the manure worm or brandling worm. This voracious eater is definitely the beginner's choice because of it's adaptability to a wide range of living conditions as well as the ability to process large amounts of organic material.

The European Nightcrawler

The European Nightcrawler is also known as *Eisenia hortensis*, the Belgium nighcrawler or super red. This composting worm is much larger than its cousin the Red Wiggler.

The African Nightcrawler

The African Nighrcrawlers is also known as *Eudrilus eugeniae*. This is a large tropical worm species that tolerates higher temperatures than its cousins listed above. This is the longest of the three composting worms and produces a lot of compost as well as breeds quickly

Where can I Raise Composting Worms?

Earthworms need only a few requirements in which to successfully live and breed. These requirements are moisture, darkness, and food. These conditions can be found in any outdoor compost pile, which is definitely the easiest type of worm system to maintain. But, because these requirments are so easily produced, many people choose to build a bin they can keep indoors, which also has its advantages.

Worms Bins (Home made / Manufactured)

Many people who are new to raising earthworms begin raising them in a Do-It-Yourself plastic storage bin or a manufactured worm bin system The convenience of portability along with the fact that most folks will not be raising populations in the tens of thousands makes the indoor worm bin system an easy project.

Composing Pile

One of the advantages of a maintaining a garden compost pile is that the finished vermicompost is conveniently close to your fruits and vegetables.

What Size Bin do I Use?

A bin that gives enough space to house your worms and bedding. For do it yourselfers, an 18 gallon (gives or take 2 gallons) bin is a good size to start with. Starting with 3 pounds of worms gives you a good amount to process your kitchen scraps.

What Kind of Bedding do I Use for Indoor Bins?

There are several types of bedding that you can use. Most popular types and their pros and cons of each type is outlined as below:

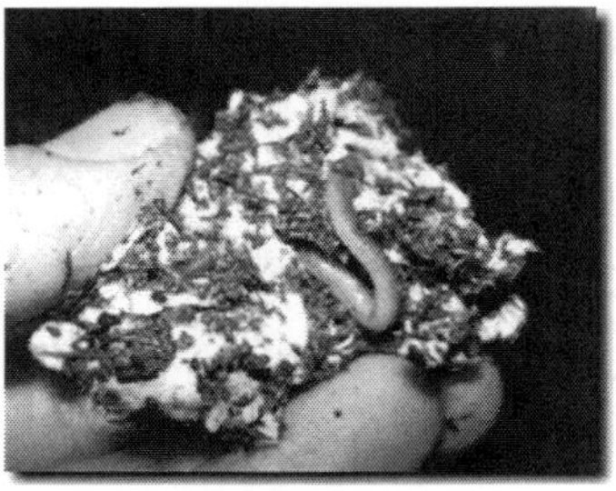

Shredded Paper

A readily available source of bedding is newspaper. You will need to shred the paper before putting it in your bin. You can hand shred it or use machine shredded paper.

Pros: Readily available, clean, no dust, odorless, easily prepared (if machine shredded). Cons: Tends to matte, making it difficult to bury waste, preparation time (if hand shredded).

Peat or Sphagnum Moss

This has long been the composter's choice for raising earthworms. It has excellent moisture retention and is readily available from your local garden or home improvement store. You will need to moisten the peat before adding to your bin. Soak until completely saturated, then squeeze until just damp like a sponge.

Manure

Manure can be used as either bedding or food and is generally used for outdoor worm beds and compost piles. You can use rabbit manure, composted cow or horse manure to top off your bin if you use peat as bedding.

Pros: Highly nutritional, often free when obtained from local stables or farms. Cons: Fresh cow and horse manure will heat up when fresh and needs to compost several weeks before use. May contain de-worming drugs.

Coconut Fibre / Coir

Coconut fiber, or coir, is a renewable material that comes primarily from India and Sri Lanka. It is becoming more popular with organic composters as well as the hydroponic growing industry. Pros: Clean, odorless, mixes well with other bedding, retains moisture well, renewable source. Cons: More expensive than other bedding material, more difficult to obtain.

It is always a good idea to "turn" or "fluff" your bedding at least once a week. You can do this by digging down to the bottom and bringing the bedding to the top. This ensures that plenty of air is in contact with the contents of your bin. You want your bedding to have plenty of oxygen so decomposition remains accelerated, plus it makes it easier for your worms to crawl through it.

Feeding - What, How and How much? What do I Feed Earthworms ?

The earthworm is an eating machine. It literally eats the earth as it burrows through it. Material that is too large to ingest will be pushed aside with its "head" (prostomium). Worms don't have teeth, so the material they eat is first moistened in their mouth then passed into their gizzard which acts like teeth and grinds the food. The food is then passed to their intestines which absorb the nutrients contained within. Whatever the worm doesn't digest is then passed out of their bodies as nitrogen rich worm castings.

An important fact to know about feeding your earthworms is that they thrive on the microscopic organisms that live on the decomposing organic matter they ingest. These organisms include a variety of algae, fungi and bacteria that are essential for the worm's growth.

When it comes to kitchen scraps, worms will consume just about everything you do, with the exception of a few.

Don'ts	*The Reason...*
Meat and Bones	Decaying meat produces offensive odor. May attract flies, rodents, ants and other non-desired pests Bones will not be processed by worms.
Salty snack food or Fast food	Big Macs, potato chips, french fries and olives are a few examples. You can soak salty foods overnight, pour off the water and then feed to your earthworms.
Pet feces	Dog or cat feces do not belong in your worm bin as they are not properly processed manure. May contain viral or bacterial toxins.
Green grass	A big mass of yard clippings will decompose thermophilically and will create high temperatures that are harmful.
Alcohol	Very toxic.
Excess citrus fruit	Citrus fruit contains a substance called limonene that can be toxic to earthworms in large quantities. You can throw in some orange peels -just not 15 oranges at once!
Pressure-treated wood	The active ingredient is cyanide which is toxic to worms in small quantities.

How do I Feed Earthworms ?

When you feed them, don't distribute the food evenly throughout the bin. Instead, put the food in the corner about halfway down into the bin. Then cover and check back in a week. You will notice that a lot of your worms will have gathered in the area you put the food! When the food is almost gone you can repeat in the opposite corner of the bin.

To speed the composting process you can chop up or boil your kitchen scraps. Raw veggies like broccoli and carrots take time to break down chopping or boiling sraps create greater surface area as well as soften and break down fibrous material, making the food more readily available to your worms.

How Much do I Feed Them ?

Just as important as what to feed worms is how much. If you have researched the internet you will find a ton of data stating that worms will eat half of their body weight daily. This is a half-truth as worms crawl throughout the soil they can ingest half of their body weight a day, but not all of what they consume will necessarily be from the kitchen scaps you are feeding them. A simple way to determine how much to give them is to make a frist note how big your fist is and put that much food in your bin. Check back in a week, note how much is left and add more when needed. You will soon get a feel for how much your worms are consuming and will be able to add more food accordingly.

A quick note about overfeeding - too much kitchen food packed together may become anaerobic and heat up your bin. If you notice that your worms seem to be trying to escape your bedding may be too hot. To cool it down, remove some of the decomposing food. Remember, the smaller your kitchen scraps are, the faster they will be consumed. Another clue that you have too much kitchen scraps is if it begins to stink. If your bin becomes smelly simply remove some of the scraps, or redistribute half to the other side of your bin.

Raising Worms for Bait

If you are raising larger worms such as European or African Nightcrawlers for fishing bait, you will want to approach growing your earthworms like a farmer. Simply feeding them your kitchen scraps won't be enough to get them as big and fat as you'll want. You'll need to feed them a steady diet of food such as manure and grains

Harvesting the Vermipost/ When to Harvest the Castings ?

After a measure of time that can be from between 1 to several months (depending on worm population density), most of the original bedding has been converted into a rich, dark earthy material. This dark material is the worm castings, also known as vermipost. When most of the original bedding has been consumed it is time to move your worms into a fresh bin of bedding.

How to Harvest the Castings ?

Choose and area with a good amount of indirect light, empty your bin onto the tarp, then separate the big pile into several smaller piles. Give the worms about an hour to crawl to the bottom of the each pile (if you do this under direct sunlight cut you wait time in half). Then carefully scoop up the "wormless" vermipost and place its own container. Now take your exposed worms and put them in a freshly bedded bin. It may be a little cumbersome at first, but you will get faster with practice. When harvesting, if there is food in your bin that has not been eaten just add it to the fresh bin.

Removing worms from the beds at regular intervals reduces overcrowding, increases production, and prevents stunted worms. Worms are initially harvested from new beds after 3 months. They can then be harvested every 2 - 4 weeks. Large quantities of worms can be snagged on the tines of a pitch fork that is slowly run through the top layer of bedding. The worms are then dropped from the pitchfork into a tub or wheelbarrow. To completely harvest a bed, a tine hoe is used to gather the top 4 inches of beddinginto a heap in the middle of the bed; the bedding and worms are transferred with a flat blade shovel into a tub, wheelbarrow, or wagon. Harvested worms are usually still covered with bedding material that clings to their bodies. This may pose a problem if the worms are to be fed to fish that require clean food and if fish tank cleaning is a problem. For small cultures, a wooden framed sieve 13 x 18 x 7 inches deep with 118 inch hardware cloth screen is helpful for separating the bedding from the worms. To rid the worms of excess bedding on a commercial scale, they are placed in an electrically-rotated basket 9 feet long x 2.5

feet diameter made of 118-inch hardware cloth stretched over a metal frame. The basket is inclined at 10-12°. As the mass of worms slides down the basket, excess bedding is scrapped away by the abrasion of the hardware cloth. The spent bedding falls into containers underneath the basket.

What is Worm Tea and How do I Make It ?

Worm tea is a solution made from worm castings and water. Think of what happens when you make tea with a tea bag; when you place a tea bag in water, the properties of the tea leaves leach into the water. The same hold true with worm castings. When you soak worm castings in water the castings are diluted and the beneficial micronutrients and microorganisms contained within are transferred to the water. This solution gives your plants and an instant boost of energy and nutrition. Not only can you feed the roots, you can use a spray bottle and feed the leaves as well. As the worm tea soaks into the soil, the nutrients and beneficial organisms are evenly dispersed into the soil and substrate. Your soil is now is teeming with molecules that your plants need to sustain healthy and productive life cycles!

Making worm tea is easy! All that you need is

Worm Castings

Water

A bucket

Worm tea bags (optional)

Simply fill the bag with castings, close the top of the bag using a bread tie and place in water to steep it's that simple.

How to Feed Earthworms to Fish ?

Earthworms are an inexpensive food for carnivorous and omnivorous fish. You can find earthworms in your back yard, purchase them from a pet store or breed them yourself. Most fish will eat earthworms but depending on the size of your fish you may need to cut the worms up to an appropriate size for the fish to be able to eat. Feeding earthworms to fish is good for their diet but is not recommended as the only item on the menu.

Instructions

1. Shake off any loose debris from the worms and wash them with water to remove dirt or pesticides that may be on their skin.
2. Cook plain oatmeal using water. Do not add additional ingredients to the oatmeal. Allow the oatmeal to cool. Once it has cooled scoop it into a container that is large enough to store the worms.

3. Place the earthworms into the container filled with oatmeal and leave them there for one day. The earthworms will eat the oatmeal and it will clean their system of dirt or pesticides that could be toxic to your fish. If the oatmeal begins to dry out, add a small amount of room temperature water to moisten it (if an earthworm's skin remains dry for too long it may die).
4. Remove the earthworms from the oatmeal after 24 hours. Rinse them off with water to remove any oatmeal that may be on their skin. Discard the oatmeal.
5. Drop a couple of earthworms, or pieces of earthworm, into your fish tank or pond. Watch for a minute or two to ensure that the earthworms are the appropriate size for your fish to eat. If the worms are too large remove them and cut them into smaller pieces. Cutting the earthworms into small pieces with sharp scissors is safer and more convenient than using a knife.
6. Repeat Steps 1 through 7 every other day or a few times a week. Do not feed earthworms to your fish every day as it is not a balanced diet for the fish.

Tips & Warnings

- Dispose of the oatmeal after 24 hours. Do not reuse it for the next batch of worms.
- Once the oatmeal has dried out it must be remoistened or replaced immediately as the worms need moisture in their environment to live -- if the worms' skin remains dry for an hour or more they may die.

Summary

Raising earthworms takes a bit of patience and care. Remember, they are living creatures that require cool, moist bedding and food they can digest. If you follow the tips given, should not have any problems. The biggest mistake that beginners tend to make is overfeeding. If you are unsure, give a little less, pay attention and feed them more when necessary.

Chapter 18

Culture of Infusorians

What is Infusoria?

The term infusoria is an old one that has changed in meaning over the years. At one time it referred to just about any microscopic, or nearly microscopic, organism that lived in fresh water.

To aquarium enthusiasts infusoria refers to many small organisms in water that tiny fry can feed up upon. Although it's not feasible to name every organism that could be in an infusoria culture, some of the primary ones include:

- Algae (Volvox); Amoebas; Euglena; Paramecium; Rotifers; Stentor; Vorticella

Thez Culture Container

Fill bottles halfway with tap water treated with a water conditioner that removes chlorine. Use treated tap water as regular tap water kills infusoria. Also use water sucked out of a planted tank.

The Culture Medium

Place one lettuce leaf in each bottle.

Infusoria Food

Place one algae wafer in each bottle, you can also try some liquid fry food.

Waiting, Waiting, and Waiting

Let both cultures sit under bright light for a week or more.

Bacteria vs Infusoria

When the water is cloudy you have bacteria but wait a little longer.

Finding Infusoria

When one culture starts clearing up or turns pink you will have infusoria. The reason it clears up is because infusoria eats bacteria.

Clone Your Best Culture

Pour your good culture into the other culture.

Feeding your fry

Feed your fry a few drops at a time so the tank stays clean. Don't over feed; it will make the water green.

Procedure

- Take plastic or glass of about 2 litres
- Add aged water from aquaria according to jar size, aquarium water, will help to develop your culture very faster
- Add 3 - 4 cabbage leaves or any other plant or animal matter
- Cover the solution with muslin cloth or cloth from fish net and leave it in a cool place away from direct sunlight and wait for 8 days
- As day's passes, you will notice the solution becoming clear and gives off less odour
- To harvest the infusoria, just take water from top 2 inches of the jar with the help of syringe and pour it in to test tube while straining it with muslin cloth to remove the unwanted plant matter
- After pouring the solution in to test tube, wait for an hour, you will see tiny organisms swimming near the surface of the water
- You can take the surface water in syringe and feed to newly born fish fry

Tips

- Siphon water from live plants as they will contain infusoria.
- Liquid fry food will jump start your cultures.
- Wash lettuce well to remove any lingering pesticides or use organic lettuce instead.

Warnings

- Don't feed until the water clears as it could kill your fry.

Chapter 19

Culture of Bloodworms

KINGDOM	:	Animalia
PHYLUM	:	Arthropoda
CLASS	:	Insecta
ORDER	:	Diptera
SUBORDER	:	Nematocera
INFRAORDER	:	Culicomorpha
SUPERFAMILY	:	Chironomoidea
FAMILY	:	Chironomidae

Bloodworms are larvae of the mosquito-like midge family Chironomidae (Order Diptera, Class Insecta). Almost all chironomids have aquatic larval and pupal stages. They have a world-wide distribution and occur in running waters (streams and rivers), standing waters (lakes and wetlands), and in waters that range from fresh to saline, from fast flowing to completely still and stagnant.

There may be more than 2000 species but only a small number have been formally identified. There is little difficulty in recognising the larvae; they are small, distinctly segmented worm-like animals. Although, they are not a true worm due to their exoskeleton and small clawed legs. Their colour is variable; some common ones are white, green, yellow, or deep red. The last is due to the presence in the larval blood of a red pigment, erythrocruorin, the presence of which is of respiratory advantage in waters with low levels of dissolved oxygen (stagnant pools). Some are also transparent and are commonly known as Glassworms. However, only those that contain erythrocruorin are red and hence the common name "Bloodworm". In some regions they are also known as "Red Mosquito Larvae".

Chironomidae larvae and pupae are highly nutritious and nourishing and constitute one of the staple food items of rainbowfishes in their natural environment.

Chemical analysis shows that bloodworms contain 9.3 % dry matter and of this there is 62.5 % crude protein, 10.4 % crude fat and 11.6 % ash with 15.4 % nitrogen free extract. They are also a good source of iron for the fish since they contain haemoglobin.

Bloodworms are a commonly used live or frozen food source for aquarium fish culture. Almost all fishes will greedily devour them when they are offered. Research has found that most fishes when provided with bloodworms as a supplementary food item have better growth and spawning rates. Frozen bloodworms could be used as a substitute for live tubificids as they have a comparable protein level.

Chironomidae go through a complete metamorphosis in their life cycle, egg, larva, pupa, and winged adult midge. Each stage having different characteristics. After mating in flight the female releases the eggs while skimming the water surface. Egg numbers can range from 50 to 700. The eggs sink to the bottom where, under tropical conditions, they hatch in 24 - 48 hours into the next stage - the larva or aquatic stage. The newly hatched larvae are not more than 1 mm long but they can measure up to 10 - 25 mm when they reach the last stage of the larva period. The larva stage can last from less than 2 weeks up to 7 weeks depending on temperature.

Each larva moults four times before it reaches the pupal stage. This stage of the chironomid forms a large part of a rainbowfishes natural diet as they leave the larval tube and actively swim to the surface of the water. Those that reach the surface emerge into flying adults after a few hours and immediately fly off to mate, living only a few hours or days. The adults do not feed during their adult existence and mating normally occurs during the night. The entire life cycle can be completed in 2 weeks, although it is common for the life cycle to take longer to complete.

Midge larvae can be found in most waters with muddy bottoms. They occur in great numbers in ponds, swamps, and streams, usually 1 - 4 metres deep. Natural breeding sites for chironomid midges are diminishing due to urbanisation, land clearing and other changes to much of the natural environment. However, they are abundant in waste water channels, sewage treatment and settlement ponds and other man-made water systems. It is these breeding areas that cause a variety of nuisance problems and public health agencies regularly spray these areas with insecticides to control their population and distribution.

If you wish to collect bloodworms the best time to catch them in large numbers is during the night when the larvae leave their self made tubes and when the dissolved oxygen at the bottom of water is low. They can be caught easily using small mesh netting. Bloodworms can also be obtained by sieving the mud on the spot. The larvae and the coarse particles of detritus will remain in the sieve and then shaken into a bucket filled with water. After a while the larvae will swim to the surface where they can be fished out with a net.

Origin	***Still bodies of water***
Maximum Size	About an inch
Life Cycle	10 - 12 days from egg to adults.
Temperature	Room temp

Origin	Still bodies of water
Water	Aerated and sponge filtered
Potential Invader	Mosquito larvae
Foods	Very fine organic matter
Lighting	Immaterial. Dark best.
Containers	Covered plastic shoe boxes
Maintenance	Daily monitoring for pupae
Harvesting	Harvest at night with a net

Blood Worms or Black Worms ? Blood worms are really the larval form of one of the chironomid midges -- insects rather than worms. You could compare them with the transparent glass worms. California blackworms, on the other hand, are a cousin to earthworms.

Origins: Blood worms grow wild in ponds, pools, and lakes. We even find their tubes in aquaria that have set unused for a period of time. Blood worms live in the mud. Glassworms swim in the water. Both can produce huge swarms.

Appeal: Fish love the taste of blood worms – even the pickiest eaters. The hemoglobin that colors them red is an excellent iron source.

Size: Full-grown blood worms grow to about an inch. Their size makes them ideal for two to six-inch fish. Once they pupate, the smaller fish sometimes have trouble eating them.

Adults: Looking very much like mosquitoes, the non-biting parents lay their eggs in water and flit off. Adults live three to five days.

Eggs: Look – under the water for blood worm eggs on sticks and grasses near the shoreline. They look like little blobs of snot with 50 to 700 tiny dots in them very much like pond snail eggs, but with lots more tiny dots. Bring these blobs home and culture them. They hatch in 24 to 48 hours.

Negative Phototropic: Blood worm larvae avoid the light. You could probably get faster growth by putting black plastic sheets over them. They construct tiny tubes in which they live during the day. This makes them difficult to harvest. The pros harvest them at night, when they emerge from their tubes. Shutting off their aeration also causes them to come out. Harvesting at night also keeps your neighborhood mosquitoes well fed.

Containers: Covered plastic shoe boxes will keep out mosquitoes. Uncovered containers are easier to aerate. Use a 10 gallon aquarium can aerate it.

Foods: Blood worms eat "micro foods." Find commercial versions or make own. Raise them on chicken manure, horse manure and other waste products. We'd normally recommend mystery snails to clean up the excess food, but you can easily imagine those rasping tongues devouring tube after tube of blood worms. Partial water changes may be necessary.

Feeding Schedule: Use 3 g / 1,000 eggs. Have fun counting the eggs. Just sprinkle some food on top. You'll probably need to monitor and feed every two to six days. This will help you keep track of their size. Feed small amounts. Overfeeding can wipe out your entire culture. Aeration helps.

Water: Good, old aquarium water or de-chlorinated tap water will work. Avoid extremes in pH levels. Their water "wears out" after four to six weeks. Their waste products "wear out" their water. That's also the reason we change the water in our fish aquariums.

Starting Out: Start with a half-dozen containers. Try different quantities of egg masses in each. Work with different foods and blood worm egg amounts. Keep notes to see what works best.

Harvesting: Net these guys out at night. Turn off your airstone and net away. Keep your excess in the fridge.

Next Generation: If you let some of your larvae pupate and turn into adults, they will lay eggs in your uncovered containers. The adults need no food. They live only to breed, lay their eggs, and continue the species.

How to Grow Bloodworms for a Fish Tank

Bloodworms are a common, nutritious and harmless live food that can be given to many aquarium fish. They are very common in lakes and ponds and can be easily captured or obtained live from aquarium suppliers and bait shops. Captive bloodworms require very little maintenance and feeding and can be grown indoors or outdoors in simple covered containers. Great quantities of bloodworms can be grown from a single batch of eggs and multiple batches can be grown concurrently with little extra difficulty.

Instructions

Keeping Bloodworms Indoors

Search for bloodworm eggs in stagnant bodies of water -- they are greenish blobs of slime with dozens to hundreds of small dot-like eggs. Check submerged portions of grasses or branches in shallow water for egg blobs.

Transfer egg blobs into a waterproof container such as a plastic shoe box or a small fish tank. Fill the container with water and cover it with a lid or dark cloth bloodworms thrive in darkness. Allow the eggs two to four days to hatch.

Provide small amounts of micro food (for freshwater organisms) or chicken or horse manure 3 g of food for every 1,000 eggs is the ideal. Feed the bloodworms every two to six days and track their size during feedings.

Perform weekly water changes to maintain water quality by replacing 10 to 20 percent of the container's volume. Store the replacement water in open containers one or two days prior to the water change this allows trapped chlorine in the water to naturally dissipate.

Harvest bloodworms when they reach a length of approximately 1 inch. Use a fine net to skim bloodworms from the water at night when they are the most active. Feed them to fish live or freeze them for long-term storage.

Use or freeze all hatched bloodworms within two weeks if they are kept indoors they will begin to pupate at this age and will eventually emerge as adult midges (flying insects). Remove any pupas that are found in the container and dispose of them outdoors.

Attracting and Keeping Bloodworms Outdoors

Bury an open waterproof container (such as a plastic bowl) up to its rim near a shading tree. Fill the container with water and place fallen leaves inside of it.

Allow the container time to attract bloodworms or place eggs or live bloodworms directly into the container. Replace evaporated water regularly. Perform weekly water changes and replace 10 to 20 % of the container's volume with dechlorinated water.

Check the water container daily and remove any mosquito larva with a fine net skimmer mosquito larva float while bloodworm larva sinks.

Harvest the bloodworms by scooping the sunken leaves onto a filter or sieve. Hold the filter or sieve against the surface of the water and allow the leaves to slowly dry the bloodworms will try to get into the water, but will be trapped by the filter. Remove the leaves and use or freeze the bloodworms.

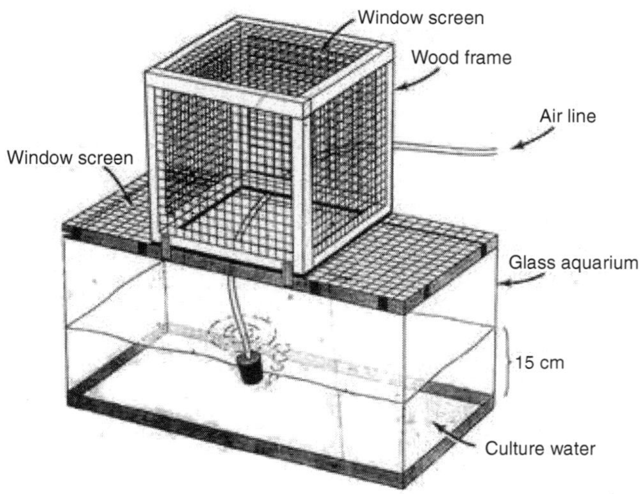

Breeding Aquarium of Chironomid larvae

Tips & Warnings

Rinse live bloodworms in water before feeding them to fish.

Adult midges are a nuisance indoors, but they are neutral or beneficial outdoors - pupating bloodworms in outdoor containers should not be removed. Place a stick in an outdoor container to incite adult midges to lay eggs in the container and quickly bolster the bloodworm population.

Chapter 20

Culture of Tubifex

KINGDOM	:	*Animalia*
PHYLUM	:	*Annelida*
CLASS	:	*Clitellata*
ORDER	:	*Oligochaeta*
FAMILY	:	*Naididae*
SUBFAMILY	:	*Tubificinae*
GENUS	:	*Tubifex (Lmarck, 1816)*

Tubifex is a cosmopolitan genus of tubificid annelids that inhabit the sediments of lakes, rivers and occasionally sewer lines. There are over ten known species of Tubifex, but the number is not certain, as the species are not easily distinguishable from each other.

Food and Feeding

Tubifex worms feed on decaying organic matter, detritus, and vegetable matter which commonly available in segment drains.

Reproduction

Tubifex worms are hermaphroditic: each individual has both male (testes) and female (ovaries) organs in the same animals. These minute reproductive organs are attached to the ventral side of the body wall in the celomic cavity. In mature specimens, the reproductive organs are clearly found on the ventral side of the body.

Copulation and Cocoon Formation

Although the Tubifex worms are hermaphrodites, the male and female organs become mature at different times; thus self-fertilization is avoided, and cross-fertilization is encouraged. Two mature Tubifex worms undergo copulation by joining their ventral and anterior surfaces together with their anterior ends pointing opposite directions. Thus, the spermathecal openings of each Tubifex worm be nearer to the male apertures of another worm. The penial setae of one worm penetrate into the tissues of other worm and thus the conjugants are held together. At this stage, the sperm of one worm is passed into the spermathecae of the other worm. After copulation, they separate and begin to produce egg cases containing eggs, called cocoons. The cocoon is formed around the clitellum as a soft, box-like structure into which the ova and the sperm are deposited. Soon the Tubifex worm withdraws its body from the egg case by its backward wriggling movements. As the worm withdraws, the cocoon completely closes and becomes a round shaped case containing both sperm cells and ova. Fertilized eggs form in the cocoon and undergo complete development in the using the case's albuminous nutritive fluid for growth. The period of development varies with temperature and lasts for two to three weeks. After complete development, the young worms emerge.

Culturing Tubifex

At least two species of Tubifex are raised commercially, mainly for fish food: the reddish Tubifex tubifex and a slightly larger, darker species, the "black worm". Tubifex can be easily cultured on mass scale in containers with 50 to 75 mm thick pond mud at the bottom, blended with decaying vegetable matter and masses of bran and bread. Continuous, mild water flow is to be maintained in the container, with a suitable drainage system. After the arrangement of the system, the container is inoculated with Tubifex worms which can be obtained from nearby muddy canals or sewage canals. Within 15 days, clusters of worms develop and can be removed with mud in masses with a spade and kept in large, wide-mouth plastic containers. When worms come to the surface due to lack of oxygen, they are collected and washed under brisk stream of water to remove residual mud attached to their bodies.

Live Food

Tubifex worms are often used as a live food for fish, especially tropical fish and certain other freshwater species. They have been a popular food for the aquarium trade almost since its inception, and gathering them from open sewers for this purpose was quite common until recently. Most are now commercially obtained from the effluent of fish hatcheries, or from professional worm farms. Using these worms as a live food has come with certain problems over the years. When harvested from sewers, open bodies of water, and even from hatcheries, they may be infected with various diseases. This risk can be partially solved by keeping the worms under brisk running water until they have voided the contents of their digestive systems. However, the Tubifex worms can still be vectors for whirling disease, which can affect salamanders and certain tropical fish. Additionally, they are very difficult for some fish to obtain in the wild, so certain fish, such as rift valley cichlids, will obsessively consume them until they make themselves sick. Additionally, while the worms have good-quality proteins, they also are very fattening, and are poor in certain important amino acids. Fish fed on tubifex can grow rapidly, but may be less healthy and colorful than fish with a more balanced diet. Lastly, in poorly cleaned aquaria, Tubifex can become established as a pest species, covering the bottom of the aquarium in a thick carpet which may be considered unsightly.

Tubifex species: The genus includes the following species:

- *Tubifex blanchardi*
- *Tubifex costatus / Tubifex ignotus*
- *Tubifex kryptus*
- *Tubifex longipenis*
- *Tubifex montanus*
- *Tubifex nerthus*
- *Tubifex newaensis*
- *Tubifex newfei*
- *Tubifex pescei*
- *Tubifex pomoricus*
- *Tubifex smirnowi*
- *Tubifex tubifex*

Chapter 21

Culture Techniques of Moina

Introduction

Daphnia are small freshwater cladoceran crustaceans commonly called "water fleas." This common name is the result not only of their size, but their short, jerky hopping movement in water. The genera *Daphnia* and *Moina* are closely related. They occur throughout the world and are collectively known as daphnia.

Daphnia have a body consisting of a head and a trunk. The antennae are the main means of locomotion. Large compound eyes lie under the skin on the sides of the head. One of the major characteristics of daphnia is that the main part of the body, the trunk, is enclosed in an external skeleton (carapace). Periodically, they molt or shed their external shell. The brood pouch, where the eggs and embryos develop, is on the dorsal side of the female. In *Daphnia*, the brood pouch is completely closed, while *Moina* have an open pouch.

There is considerable size variation between the genera. *Moina* are approximately half the maximum length of *Daphnia*. Adult *Moina* (700 - 1,000 µm) are longer than newly-hatched brine shrimp (500 µm) and approximately two to three times the length of adult rotifers. Young *Moina* (< 400 µm), however, are approximately the same size or only slightly larger than adult rotifers and smaller than newly-hatched brine shrimp. In addition, brine shrimp die quickly in freshwater. As a result, *Moina* are ideally suited for feeding freshwater fish fry.

Newly hatched fry of most freshwater fishspecies can ingest young *Moina* as their initial food. However, it should be noted that it can be difficult to grade *Moina* for size. It was found through trials at

Tthe UF/IFAS Tropical Aquaculture Laboratory that passing *Moina* through 500 micron meshes screeningtends to fragment the animals to such an extent that they are no longer usable as live food. In aquaria, care ust also be taken when determining feeding rates, as *Moina* can quickly grow too large to be eaten. If these larger *Moina* become too dense, their "hopping" movements can serve to harass and potentially damage fry. In Singapore, *Moina micrura* grown in ponds, fertilized with mostly chicken manure or, less frequently, with pig manure, are used as the sole food for fry of many ornamental tropical fish species, with a 95 - 99 % survival rate to 3/4 inch (20 mm) in length quite common. Unfortunately, there is very little information concerning practical mass culture methods of *Moina*, and the available information is in mimeograph documents, foreign journals or other scarce publications.

Physical and Chemical Requirements

Moina appear in high concentrations in pools, ponds, lakes, ditches, slow-moving streams andswamps where organic material is decomposing. They become especially abundant in temporary water bodies which provide them with suitable conditions for only a brief period. *Moina* are generally quite tolerant of poor water quality. They live in water where the amount of dissolved oxygen varies from almost zero to supersaturation. *Moina* are particularly resistant to changes in the oxygen concentration and often reproduce in large quantities in water bodies strongly polluted with sewage. Species of *Moina* have been reported to play an important role in the stabilization of sewage in oxidation lagoons. The ability to survive in oxygen-poor environments is due to their capacity to synthesize hemoglobin. Hemoglobin formation is dependent on the level of dissolved oxygen in the water. The production of hemoglobin may also be caused by high temperature and high population density. *Moina* are resistant to extremes in temperature and easily withstand a daily variation of 41 - 88° F (5 - 31° C); their optimum temperature is 75 - 88° F (24 - 31° C). The high temperature tolerance of *Moina* is of great advantage for both the commercial fish farmers in the southern U.S. and hobbyists culturing live food at home.

Food Requirements

Moina feed on various groups of bacteria, yeast, phytoplankton and detritus (decaying organic matter). Bacterial and fungal cells rank high in food value. Populations of *Moina* grow most rapidly in the presence of adequate amounts of bacterial and yeast cells as well as phytoplankton. *Moina* are one of the few zooplankton which can utilize the blue-green algae *Microcystis aeruginosa*. Both plant and animal detritus may provide energy for the growth and reproduction of *Moina*. The food value of detritus depends on its origin and diminishes with the age of the detritus.

Life Cycles of Moina

The reproductive cycle of *Moina* has both a sexual and asexual phase. Normally, the population consists of all females that are reproducing asexually. Under optimum conditions, *Moina* reproduce at only 4 - 7 days of age, with a brood size of 4 - 22 / female. Broods are produced every 1.5 - 2 days, with most females producing 2 - 6 broods during their lifetime. Under adverse environmental conditions, males are produced and sexual reproduction occurs resulting in resting eggs (ephippia), similar to brine shrimp eggs. The stimuli for the switch from asexual to sexual reproduction in populations of *Moina* are an abrupt reduction in the food supply, resulting in an increase in resting egg production. However, it is advantageous to keep the population well fed and in the asexual mode of reproduction, since fewer progeny are produced with resting eggs. High population densities of *Daphnia* can result in a dramatic decrease in reproduction, but this is apparently not the case with *Moina.* The egg output of *Daphnia magna* drops sharply at a density as low as 95-115 mature individuals per gallon (25 - 30 / L). The maximum sustained density in cultures of *Daphnia* reported is 1,900 individuals / gallon (500 / L). *Moina* cultures, however, routinely reach densities of 19,000 individuals per gallon (5,000 / L) and are, therefore, better adapted for intensive culture. A comparison of the production of *Daphnia magna* and *Moina macrocopa* cultures fertilized with yeast and ammonium nitrate, showed that the average daily yield of *Moina* (1.42 - 1.47 ounces / 100 gallons; 106 - 110 g / m^3) is three to four times the daily production of *Daphnia* (0.33 - 0.53 ounces / 100 gallons; 25 - 40 g / m^3). The daily yield of *Moina* cultures fed phytoplankton cultured on organic fertilizer have been reported to exceed 5 ounces/100 gallons (375 g / m^3).

Nutritional Value of Moina

The nutritional content of *Moina* varies considerably depending on their age and the type of food they are receiving. Although variable, the protein content of *Moina* usually averages 50% of the dry weight. Adults normally have a higher fat content than juveniles. The total amount of fat per dry weight is 20 - 27 % for adult females and 4 - 6 % for juveniles.

Procedure for Moina Culture

The batch culture method of producing *Moina* uses a continuous series of cultures. Briefly, a new culture is started daily in a separate container using the procedures outlined below. When all the fungal, bacterial, and algal cells are consumed, usually about 5 - 10 days after inoculation, the *Moina* are completely harvested, and the culture is restarted. This method is particularly applicable when a specific quantity of *Moina* is needed each day because daily production is much more controlled. Batch culture is also useful for maintaining pure cultures because there is less chance of the cultures becoming contaminated with competitors (e.g.,protozoans, rotifers, copepods) or predators of fish larvae or fry (e.g., *Hydra,* back-swimmers, diving beetles, dragonfly larvae). Semi-continuous cultures can be maintained for two months or more by daily partial harvests of *Moina,* water changes and regular feeding, keeping the population in a state of rapid growth. Eventually, the *Moina* cultures will fail to respond to additional fertilization. When it is evident that they are not reproducing well, the *Moina* should be completely harvested and a new culture started. *Moina* can be produced either in combination with their food or as separate cultures. Combined culture is the simplest, but production from separate cultures has been reported to be approximately higher. For separate culture, the phytoplankton tank is positioned so that it can be drained into the *Moina* culture tank (Figure 2). Production from separate cultures has the disadvantage of requiring additional space for the cultivation of phytoplankton. However, there are advantages of separate culture of *Moina* and phytoplankton. The advantanges include less chance of contamination, a greater degree of control, and more consistent yield.

Containers

Cultures have been maintained in 10 gallon (38 L) aquaria. However, this volume usually, only yields enough *Moina* for the hobbyist culturing live food. For larger scale and commercial operations, tanks or vats (concrete, stainless steel, plastic or fiberglass) and earthen ponds can be used. Wading pools, plastic sinks, old bathtubs, discarded refrigerator liners and cattle watering troughs also work well. Do not use unpainted metal containers unless they are stainless steel. In these larger containers, water depth should be no greater than 3 feet (0.9 m). A maximum depth of 16 - 20 inches (0.4 - 0.5 m) is probably optimum. The shallow water depth allows good light penetration for photosynthesis by phytoplankton and provides a large surface to volume ratio for oxygen diffusion. Diffuse light or shade over of the water surface of the *Moina* culture container is recommended. A greenhouse covered with shade cloth (50 - 80 % light eduction) is ideal. Outdoor cultures should be protected from rain to help stabilize production and screened to prevent entry of predacious aquatic insects. Containers to be used, whether aquaria, tanks,vats or ponds, need not be particularly clean. However, filamentous algae and predators of fish larvae or fry (e.g., *Hydra*, back swimmers, diving beetles, dragonfly larvae) can be especially troublesome in *Moina* cultures. Tanks can be disinfected with a 30 % solution of muriatic acid or by drying in sunlight. Earthen ponds should be drained and sun dried.

Water

Moina are extremely sensitive to pesticides, metals (e.g., copper and zinc, which may be prevalentin municipal or well water), detergents, bleaches and other toxic materials in the water supply. Ensure that toxins are not inadvertently introduced into the culture container. Well water should be aerated for at least two hours. Municipal water should be aerated for at least two days to neutralize the chlorine, or sodium thiosulfate or a commercially available chlorine neutralizer can be added to shorten this process. Natural spring water is ideal. Rain water is also excellent for *Moina* cultures if it is collected from an area that does not have excessive air pollution. Filtered lake or stream water may also be used. The optimum water temperature for *Moina* is 75 - 88° F (24 - 31° C). *Moina* continue to thrive at temperatures in excess of 90° F (32° C) for short periods. However, low temperatures reduce production.

Aeration

Gentle aeration of the *Moina* pool oxygenates the water, keeps food particles in suspension andincreases phytoplankton production; these result in an increase in the number of eggs per female, the proportion of egg bearing females in the population, and the population density. A small trickle of freshwater into the culture container may also improve production of *Moina*. Only one or two aquarium air lines are required in culture containers up to 400 gallons (1.5 m^3). Extremely small bubbles should beavoided as they can get trapped under the carapace, causing *Moina* to float at the surface, eventually killing them.

Feeding or Fertilizing

Listed below are some common fertilizer materials and application rates. Try several of these culture media to determine which one works best in your situation. The initial fertilization rates provided are only a starting point and will probably need to be adjusted depending on individual culture conditions.The following quantity of fertilizer materials should be added initially for each 100 gallons (379 L) of water. Additional feed or fertilizer, approximately 50 - 100 % of the initial amount, should be added about 5 days later.

1. Yeast: 0.3 - 0.5 ounces (8.5 - 14.2 g) of baker's yeast
2. Yeast and mineral fertilizer: 0.3 - 0.5 ounces (8.5 - 14.2 g) of yeast and 0.5 ounces (14.2 g) of ammonium nitrate
3. Alfalfa, bran and yeast: 1.5 ounces (42.5 g) of alfalfa pellets or meal, 1.5 ounces (42.5 g) of wheat or rice bran, and 0.3 ounces (8.5 g) of yeast
4. Cow manure or sewage sludge, bran and yeast: 5 ounces (142 g) of dried manure or sewage sludge, 1.5 ounces (42.5 g) of wheat or rice bran, and 0.3 ounces (8.5 g) of yeast
5. Cow manure or sewage sludge, cotton seed meal and yeast: Use 5 ounces (142 g) of dried manure or sewage sludge, 1.5 ounces (42.5 g) of cotton seed meal and 0.3 ounces (8.5 g) of yeast
6. Horse or cow manure or sewage sludge: Combine 20 ounces (567 g) of dried manure or sewage sludge
7. Chicken or hog manure: Combine 6 ounces (170 g) of dried manure
8. Yeast and spirulina powder: 0.2 ounces (6 g) bakers yeast, 0.1 ounces (3 g) spirulina powder

Add this amount for the first two days, and then every other day until culture is harvested. ***Note:*** Add warm water to yeast and spirulina powder and let sit for 30 minutes. Stir and pour contents through a brine shrimp net into the *Moina* culture. The net will catch un-dissolved yeast and extend the life of the culture. Organic fertilizers are usually preferred to mineral fertilizers because organic fertilizers provide bacterial and fungal cells and detritus as well as phytoplankton as food for the *Moina*. This variety of food items more completely meets their nutritional needs, resulting in maximum production. Mineral fertilizers may be used alone, however, they work better in earthen ponds than in tanks or vats.Fresh manures are preferred because they are richer in organic matter and bacteria. However, some farm nimals are given feed additives that control fly larvae in their manure and these may inhibit the production of *Moina*. Although not absolutely necessary, the manure is frequently dried before use. Commercially available organic fertilizers, such as dehydrated cow manure and sewage sludge, may beused for *Moina* cultures. Although manure is widely used to culture *Moina*, yeast, alfalfa and bran are less objectionable to use and they work well. Activated yeast (baker's yeast) is readily available from wholesale food distributors in 2 pound (0.9 kg) bags. Bran and alfalfa meal or pellets can be purchased in 50 pound (22.7 kg) bags from livestock feed stores. Coarse organic materials, such as manure, sewage sludge, hay, bran and oil seed meals, are usually suspended in the water column in mesh bags. Cheese cloth, burlap, muslin, nylon or other relatively loose weave fabrics may be used. Nylon and other synthetic fabrics, however, do not deteriorate in water as do cotton or burlap. For smaller culture containers, nylon stockings work well for this purpose, are inexpensive and readily available. The use of a bag prevents large particles from being a problem when the *Moina* are harvested and allows greater control of fertilization. Over feeding can quickly cause problems in water quality. Regardless of the type of media used, start with small amounts of feed or fertilizer added at frequent intervals and slowly increase the amount used as you gain experience. If fungus occurs in the culture container due to over fertilization, the bag containing the organic material should be removed from the culture. If fungus persists in large quantities the culture should be discarded and restarted. Excessively high pH (greater than 9.5), due to a heavy algae bloom and the resulting increase in the proportion of the toxic form of ammonia (un-ionized), may inhibit the production of *Moina*. The pH of the culture can be adjusted to 7 - 8 with vinegar (acetic acid).

Inoculating

Use pure live cultures to inoculate. Avoid using animals for inoculation from poor or declining cultures, cultures producing resting eggs, or cultures containing predators of fish larvae or fry. Inoculate with approximately 100 *Moina* gallon (25 / L). Although a culture can theoretically be started with a single female, always use an adequate number to develop a harvestable population quickly. If fewer are used, the population in the culture will increase more slowly, therefore, the initial quantity of fertilizer or food should be reduced to prevent over feeding. A greater number used for inoculation reduces the time to harvesting and lessens the chance of contamination by competitors. Cultures are usually inoculated 24 hours or more after fertilization. However, when yeast is used, *Moina* can be added to the culture after a few hours of aeration, assuming good water quality and proper temperature. This is because the yeast cells are immediately available to the *Moina* as food. The small amount of phytoplankton present in the water and digestive tract of the *Moina* used to inoculate the culture is usually sufficient to initiate a phytoplankton bloom.

Monitoring

The culture should be inspected daily to determine its health. The following observations should be made.

1. The health of the culture is determined by stirring the culture, removing 1 tablespoon (15 ml) of the culture, and examining the sample with an 8 X to 10 X hand lens or dissecting scope. Green or brown red *Moina* with full intestinal tracts and active movement indicate a healthy culture. Pale *Moina* with empty digestive tracts or *Moina* producing resting eggs are indications of suboptimum environmental conditions or insufficient food.
2. The population density of *Moina* is determined by killing the *Moina* in a 1 teaspoon (3 - 5 / ml) sample with a 70 % alcohol solution and counting all

Moina in a petri dish with a hand lens or dissecting scope. Cultures ready for harvest should contain 45 - 75 *Moina* in the 1 teaspoon sample (3 - 5 / ml). With experience, population density can be estimated visually without the need for counts.

3. The food concentration in the culture water, when examined in a clear glass, should appear slightly cloudy and tea colored or green. Clear culture water is an indication of insufficient food. The cultures should be fed or fertilized with approximately 50 - 100 % of the initial quantity whenever the transparency is greater than about 12 - 16 inches (0.3 - 0.4 m). This can be determined with a white plastic or metal lidapproximately 4 inches (100 mm) in diameter, attached to the end of a yard stick. The depth of transparency is where the disk is just barely visible when lowered into the tank.
4. If predators of fish larvae or fry (e.g., *Hydra*, back swimmers, diving beetles, dragonfly larvae) are observed, discard the culture and clean and disinfect the tank or pool to avoid contaminating other cultures.

Harvesting

Moina can be harvested by simply dipping out the required number with a brine shrimp net or re-usable coffee filter as they concentrate in "clouds" at the surface. Cultures may also be harvested by draining or siphoning the culture water into a plankton collector equipped with 50 to 150 μm mesh netting net suspended in a container of water. Turn off the aeration and allow the food particles to settle before harvesting. For semi-continuous culture, do not harvest more than 20 - 25 % of the population each day, unless you are restarting the culture. Harvesting by draining the culture tank allows for a partial water exchange, improving water quality. Harvest only small quantities at a time and transfer the *Moina* to containers with fresh water to keep them alive. The bottom sediments should be stirred up manually every day following harvest, thoroughly mixing the culture to re-suspend food particles and prevent anaerobic conditions from developing.

Additional Points

Differences in size, brood production and optimum environmental conditions exist between different species and varieties of *Moina*. Adjustments will need to be made in the culture technique depending on the particular species or variety you wish to produce. Additional surfaces in the culture tank may have a positive effect on the production of *Moina*. For *Daphnia*, a four fold increase of surface area, in the form of plastic sheets, has been shown to result in a four-fold increase in the density, biomass and harvest. It is unknown whether this is the result of improved water quality due to nitrifying bacteria on the substrate, a change in the spatial distribution of the *Daphnia* or improved nutrition. It may not always be possible to match *Moina* production to the food demand of the fish fry. Harvested *Moina* can be kept alive for several days in clean water in a refrigerator. They will resume normal activity when they are again warmed. The nutritional quality of the stored *Moina* will probably not be optimal because of the period of starvation, so the *Moina* should be enriched with algae and yeast before feeding them to fish. *Moina* can be stored for long periods by freezing in low salinity water (7 ppt, 1.0046 densities) or by freeze-drying. Both methods kill the *Moina,* so adequate circulation is required to keep them in suspension after thawing so they will be available to the fish fry. Frozen and freeze-dried *Moina* are not as nutritious as live animals and they are not as readily accepted by fish fry. Although freezing or freeze-drying does not significantly alter the nutritional content of *Moina,* nutrients do leach out rapidly into the water. Nearly all of the enzyme activity is lost within ten minutes after introduction in fresh water. After one hour, all of the free amino acids and many of the bound amino acids are lost.

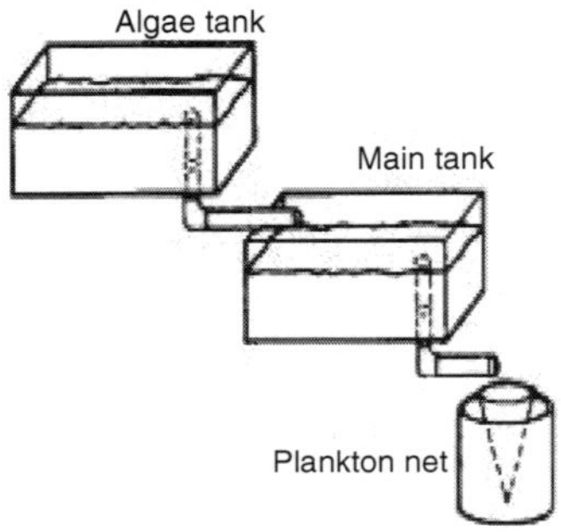

Tank arrangement for separate culture of *Moina* & its food

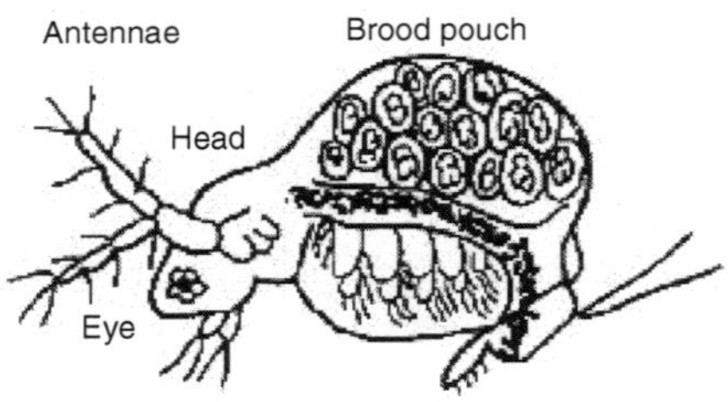

In *Daphnia*, the brood pouch is completely closed, while *Moina* have an open pouch Adult *Moina*

Chapter 22

Culture of Daphnia

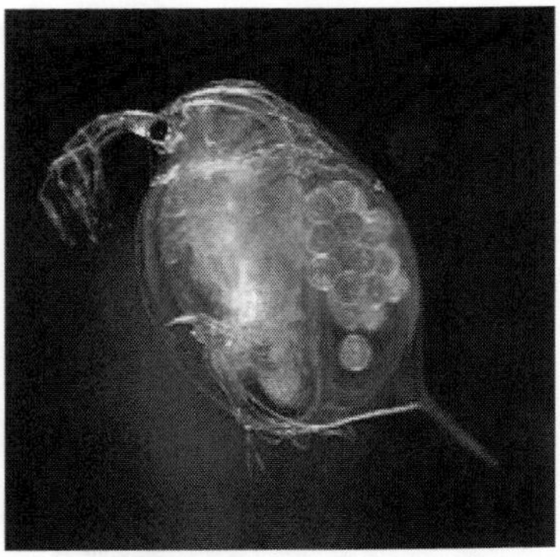

Daphnia Magna - Female Adult

Cladocera is an order of small crustaceans commonly called water fleas. Around 620 species have been recognised so far, with many more undescribed. They are ubiquitous in inland aquatic habitats, but rare in the oceans. Most are 0.2 - 6.0 mm (0.0079 - 0.24 inch) long, with a down-turned head, and a carapace covering the apparently unsegmented thorax and abdomen. There is a single median compound eye. Most species show cyclical parthenogenesis, where asexual reproduction is occasionally supplemented by sexual reproduction, which produces resting eggs that allow the species to survive harsh conditions and disperse to distant habitats.

KINGDOM	:	*Animalia*
PHYLUM	:	*Arthropoda*
SUBPHYLUM	:	*Crustacea*
CLASS	:	*Branchiopoda*
ORDER	:	*Cladocera*
FAMILY	:	*Daphniidae*
GENUS	:	*Daphnia*
SPECIES	:	*D. magna*

Daphnia magna (Straus, 1820)

Biology and Life Cycle of Daphnia

Daphnia is a frequently used food source in the freshwater larviculture (e.g., different carp species) and in the ornamental fish industry (e.g., guppies, sword tails, black mollies and plattys). Daphnia belongs to the suborder Cladocera, which are small crustaceans that are almost exclusively living in freshwater. The carapace encloses the whole trunk, except the head and the apical spine (when present). The head projects ventrally and somewhat posteriorly in a beak-like snout. The trunk appendages (five or six pairs) are flattened, leaf-like structures that serve for suspension feeding (filter feeders) and for locomotion. The anterior part of the trunk, the post abdomen is turned ventrally and forward and bears special claws and spines to clean the carapace. Species of the genus Daphnia are found from the tropics to the arctic, in habitats varying in size from small ponds to large freshwater lakes. At present 50 species of Daphnia are reported worldwide, only six of them normally occur in tropical lowlands.

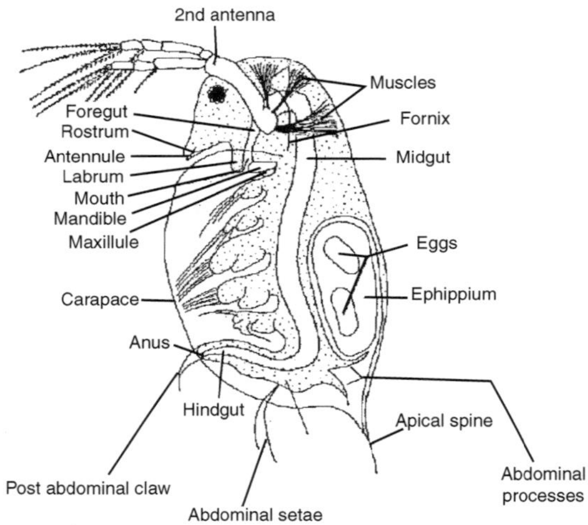

Schematic drawing of internal & external anatomy of Daphnia

The adult size is subjected to large variations; when food is abundant, growth continues throughout life and large adults may have a carapace length twice that of newly-mature individuals. Apart from differences in size, the relative size of the head may change progressively from a round to helmet-like shape between spring and midsummer. From midsummer to fall the head changes back to the normal round shape. These different forms are called cyclomorphs and may be induced, like in rotifers, by internal factors, or may be the result from an interaction between genetic and environmental conditions.

Normally, there are 4 to 6 Instar stages. Daphnia grows from nauplius to maturation through a series of 4 - 5 molts, and the period depends primarily on temperature (11 days at 10°C to 2 days at 25°C) and the availability of food. Daphnia species reproduce either by cyclical or obligate parthenogenesis and populations are almost exclusively female. Eggs are produced in clutches of two to several hundred, and one female may produce several clutches, linked with the molting process. Parthogenetic eggs are produced ameiotically and result in females, but in some cases males can appear. In this way the reproductive pattern is similar to rotifers, where normally parthenogenetic diploid eggs are produced. The parthenogenetic eggs (their number can vary from 1 to 300 and depends largely upon the size of the female and the food intake) are laid in the brood chamber short after ecdysis and hatch just before the next ecdysis. Embryonic development in cladocerans occurs in the broodpouch and the larvae are miniature versions of the adults. In some cases the embryonic period does not correspond with the brood period, this means that the larvae are held in the brood chamber even after the embryonic period is completed, due to postponed ecdysis (environmental factors). For different species the maturation period is remarkably uniform at given temperatures: 11 days at 10°C and only 2 days at 25°C. Factors, such as change in water temperature or food depreviation as a result of population increase, may induce the production of males. These males have one or two gonopores, which open near the anus and may be modified into a copulatory organ. The male clasps the female with the first antennae and inserts the copulatory processes into the single, median female gonopore. The fertilized eggs are large, only two are produced in a single clutch (one from each ovary), and are thick-shelled: these resting or dormant eggs are enclosed by several protective membranes, the ephippium. They are resistant to dessication, freezing and digestive enzymes; in this respect they can play an important role in colonizing new habitats or in the re-establishment of an extinguished population after unfavourable seasonal conditions.

Nutritional Value of Daphnia

The nutritional value of Daphnia depends strongly on the chemical composition of their food source. But, since Daphnia is a freshwater species, it is not a suitable prey organism for marine organisms, due to its low content in essential fatty acids: i.e. (n-3) HUFA. Furthermore, Daphnia contains a broad spectrum of digestive enzymes such, as proteinases, peptidases, amylases, lipases and even cellulose that can serve as exo-enzymes in the gut of the fish larvae.

Feeding and Nutrition of Daphnia

The filtering apparatus of Daphnia is constructed of specialized thoracic appendages for the collection of food particles. Five thoracic limbs are acting as a suction and pressure pump. The third and fourth pair of appendages carries large filter-like screens which filter the particles from the water. The efficiency of the filter allows even the uptake of bacteria (approx. 1micro meter). Heterotrophic microflagellates and ciliates up to the size of Paramecium can also be used as food for Daphnia. Even detritus and benthic food can be an important food source, especially when the food concentration falls below a certain threshold. In this case, the water current produced by the animals swimming on the bottom whirls up the material which is eventually ingested. Since daphnids seem to be non-selective filter feeders (i.e., they do not discriminate between individual food particles by taste) high concentrations of suspended material can interfere with the uptake of food particles.

General Procedure for Tank Culture

Daphnia is very sensitive to contaminants, including leaching components from holding facilities. When plastic or other polymer containers are used, a certain leaching period will be necessary to eliminate toxic compounds. The optimal ionic composition of the culture medium for Daphnia is unknown, but the use of hard water, containing about 250 mg / l of CO_3, is recommended. Potassium and magnesium levels should be kept under 390 mg / l and 30 - 240 g / l, respectively. Maintenance of pH between 7 to 8 appears to be important to successful Daphnia culture. To maintain the water hardness and high pH levels, lime is normally added to the tanks. The optimal culture temperature is about 25^{0}C and the tank should be gently aerated to keep oxygen levels above 3.5 mg / l (dissolved oxygen levels below 1.0 mg / l are lethal to Daphnia). ammonia levels must be kept below 0.2 mg / l. Inoculation is done using adult Daphnia or resting eggs. The initial density is generally in the order of 20 to 100 animals per liter. Normally, optimal algal densities for Daphnia culture are about 10^5 to 10^6 cells / ml (larger species of Daphnia can support 10^7 to 10^9 cells / ml). There are two techniques to obtain the required algal densities: the detrital system and the autotrophic system.

Detrital System

The "stable tea" rearing system is a culture medium made up of a mixture of soil, manure and water. The manure acts as a fertilizer to promote algal blooms on which the daphnids feed. One can make use of fresh horse manure (200 g) that is mixed with sandy loam or garden soil (1 kg) in 10 l pond water to a stable stock solution; this solution diluted two to four times can then be used as culture medium. Other fertilizers commonly used are: poultry manure (4 g / l) or cow-dung substrates. This system has the advantage to be self-maintaining and the Daphnia are not quickly subjected to deficiencies, due to the broad spectrum of blooming algae. However, the culture parameters in a detrital system are not reliable enough to culture Daphnia under standard conditions, e.g. over fertilization may occur, resulting in anoxic conditions and consequently in high mortalities and/or ephippial production.

Autothrophic System

Autotrophic systems on the other hand use the addition of cultured algae. Green water cultures (10^5 to 10^6 cells / ml) obtained from fish pond effluents are frequently used but these systems show much variation in production rate mainly because of the variable composition of algal species from one effluent to another. Best control over the culture medium is obtained when using pure algal cultures. These can be monocultures of e.g. Chlorella, Chlamydomonas or Scenedesmus, or mixtures of two algal cultures. The problem with these selected media is that they are not able to sustain many Daphnia generations without the addition of extra vitamins to the Daphnia cultures.

To calculate the daily algal requirements and to estimate the harvesting time, regular sampling of the population density must be done. Harvesting techniques can be non-selective irrespective of size or age group or selective (only the medium sized daphnids are harvested, leaving the neonates and matured individuals in the culture tank).

Mass cultivation of Daphnia magna can also be achieved on cheap agro-industrial residues, like cotton seed meal (17 g /l), wheat bran (6.7 g / l), etc. Rice bran has many advantages in comparison to other live foods (e.g. microalgae): it is always available in large quantities; it can be purchased easily at low prices; it can be used directly after simple treatment (micronisation, defatting); it can be stored for longer periods; it is easy to dose; and it has none of the problems involved in maintenance of algal stocks and cultures.

Next to these advantages, there is also the fact that rice bran has a high nutritional value. Rice bran (defatted) contains 24 % (18.3 %) crude protein, 22.8 % (1.8 %) crude fat, 9.2 % (10.8 %) crude fiber, and is rich in vitamins and minerals. Daphnia can be grown on this food for an unlimited number of generations without noticeable deficiencies.

Defatted rice bran is preferred above raw rice bran, because it prevents hydrolysis of the fatty acids and, consequently, rancidity of the product. Micronisation of the bran into particles of less than 60 m is done by treating an aqueous suspension (50 g / l) with a handmixer and filtering it through a 60 m sieve, or by preparing it industrially by a dry mill process. The suspension is administered in small amounts throughout a 24 h period: 1 g of defatted rice bran per 500 individuals for two days (density: 100 animals / l). The food conversion ratio has an average of 1.7, which means that with less than 2 kg of dry rice bran approximately 1 kg wet daphnid material can be produced (with a 25 % water renewal per week) (De Pauw *et al.*, 1981).

General Procedure for Pond Culture

Daphnia can also be produced in ponds of at least 60 cm in height. To produce 1 ton of Daphnia biomass per week, a 2500 m^3 culture pond is required. The pond is filled with 5 cm of sun-dried (for 3 days) soil to which lime powder at a rate of 0.2 kg lime powder per ton soil is added. After this the pond is filled with water up to 15 cm height. Poultry manure is added to the ponds on the 4th day at a rate of 0.4 kg / m^3 to promote phytoplankton blooms. Fertilization of the pond with organic manure instead of mineral fertilizers is preferred because cladocerans can utilize much of the manure directly under the form of detritus. On day 12 the water level is raised to 50 cm and the pond is for the second time fertilized with poultry manure (1 kg / m^3). Weekly fertilization rates are maintained at 4 kg poultry manure per m^{-3}. Also fresh cow dung may be used: herefore a suspension is made up containing 10 g / l, which is filtered through a 100 m sieve. In the first week 10 l extract per day and per ton of water is used. This ratio is increased during the following weeks from 20 l / m^3 /day in the 2nd week to 30 l / m^3 / day in the following weeks.

Inoculation of the ponds is done on the 15th day at a rate of 10 daphnids per liter. One month after the inoculation, blooms of more than 100 g / m^3 can be expected. To maintain water quality in these ponds, fresh hard water can be added at a maximum rate of 25 % per day. Harvesting is done by concentrating the daphnids onto a 500 m sieve.

The harvested biomass is concentrated in an aerated container (< 200 daphnids / l). In order to separate the daphnids from unfed substrates, exuviae and faecal material, the content of the container is brought onto a sieve, which is provided with a continuous circular water flow. The unfed particles, exuviae and faeces will collect in the centre on the bottom of the sieve, while the daphnids remain in the water column. The unwanted material can then be removed by using a pipette or sucking pump. Harvesting can be complete or partial; for partial harvesting a maximum of 30 % of the standing crop may be harvested daily.

Contamination

Daphnia cultures are often accidentally contaminated with rotifers. Especially Brachionus, Conochilus and some bdelloids may be harmfull, e.g. B. rubens lives on the daphnids and causes a hindrance in swimming and food collection activities. Brachionus is simply removed from the culture by flushing the water and using a sieve of appropriate mesh size as Daphnia is much bigger than Brachionus. Conochilus, on the other hand, can be eliminated by adding cow dung to the culture (lowering the oxygen levels). Bdelloids are more difficult to remove from the culture since they are resistant to a wide range of environmental conditions or even drought. Elimination is possible by creating strong water movements, which bring the bdelloids (which are bottom dwellers) in the water column, and remove them by using sieves.

Production and Use of Resting Eggs

Resting eggs are interesting material for storage, shipment and starting of new Daphnia cultures. The production of resting eggs can be initiated by exposing a part of the Daphnia culture to a combination of stressful conditions, such as low food availability, crowding of the animals, lower temperatures and short photoperiods. These conditions are generally obtained with aging populations at the end of the season. Collection of the ephippia from the wild can be done by taking sediment samples, rinsing them through a 200 m sieve and isolating the ephippia under a binocular microscope. Normally, these embryos remain in dormancy and require a diapause inhibition to terminate this status, so that they can hatch when conditions are optimal. Possible diapause termination techniques are exposing the ephippia to low temperatures, darkness, oxygen and high carbon dioxide concentrations for a minimal period of several weeks (Davison, 1969).

There is still no standard hatching procedure. Generally the hatching is stimulated by exposing the ephippia to higher temperatures (17 - 24^{0}C), bright white light (70 W / m^2), longer photoperiods and high levels of dissolved oxygen. It is important, however, that these shocks are given while the resting eggs are still in the ephippium. After the shock the eggs may be removed from the ephippium. The hatching will then take place after 1 - 14 days.

Chapter 23

Bioenrichment of Artemia nauplii

The process of incorporating nutrients or medicines into living organism that can then be fed to the target fish, e.g. polyunsaturated fatty acids, important in early larval development, can be encapsulated in rotifers for feeding to marine fish larvae.

Factors to be considered prior to bioencapsulation

i. Selection of the carrier or the live food, taking into account the acceptability of the organism and its size. Commonly used carriers are:

   a) Micro algae (2 - 20)

   b) Rotifers (5 - 200)

   c) Artemia (20D - 400)

   d) Moina (400 - 1500)

   e) Daphnia (20D - 400)

ii. Nutritional quality, digestibility and acceptability before and after encapsulation

iii. Fixing up of the level of enrichment media to be incorporated into the carrier organism, which depends on the nutritional quality of the carrier organism before. Incorporation and based on feeding trials conducted in laboratory.

iv. Economic feasibility of enrichment

v. Purity of the culture of carrier organism and

vi. The easy procurement of carrier organisms, its viable culture techniques; catchability of carrier organism by target species, and its easy reproduction. If the entire are satisfied, one can go for a effective enrichment.

Enrichment with highly unsaturated fatty acids (HUFA)

Enrichment of Artemla nauplil with free amino acids (FAA)

Bioencapsulation with vitamin C

Bjpencapsulation with drugs

Simplified formula for Artemia enrichment

Ingredients for enrichment

Gelatin 30 g (unflavored)

Boiled deionized water 800 ml

Cod liver oil 160 ml or other high omega fish oil

Raw egg yolks 4 (Egg yolks can be substituted with soybean lecithin 8 g or 5 % of the weight of the oil)

Vitamin premix, including E, C and B complex 10 g

Beta-carotene 1 g

Procedure

Dissolve gelatin in 800 ml boiled deionized water and let it cool to 40°C. Mix the oil in a blender on the highest setting for 30 seconds while adding beta carotene. With blender on, add vitamins and egg yolks. Then add gelatin and blend for 90 seconds. Store the product covered in the refrigerator. Enrichment media along with hatching process Use about 0.5 ml of the enrichment diet per liter of incubation water (assuming 2 g dry cysts per liter of incubation water). After 18 to 24 hours hatching time, add another 0.5 ml of the enrichment diet per liter 2 hours before harvesting. Harvest before Artemia become too large for fish larvae. Enrichment of separated Artemia Use about 0.5 ml of the enrichment diet per liter of separated Artemia (assuming a density of 100 to 150 Artemia / ml) for not less than 4 hours. Aerate the Artemia / diet mix during the enrichment process. Cooling the water with ice may slow the rate at which the Artemia grow during the enrichment process if size is critical to the target larvae. The drawback to this procedure is that Artemia grow very rapidly, metamorphose to second instar metanauplii, and can become too large for some fish larvae to consume.

Artemia are being used frequently as a food for larval fish. It has been demonstrated that lack of n-3 Highly Unsaturated Fatty Acids (HUFA) such as Eicosapentaenoic acid (EPA) and Docosahexaenoic acid (DHA) and also the DHA/ EPA-ratio, is the principal factor affecting the dietary value of Artemia. Therefore, Artemia must be always enriched with oils containing high levels of n-3 HUFA. The technique is called "Boosting". This chapter explains the essentiality of Artemia enrichment, the various techniques involved in this regard and in the preparation of emulsion etc.

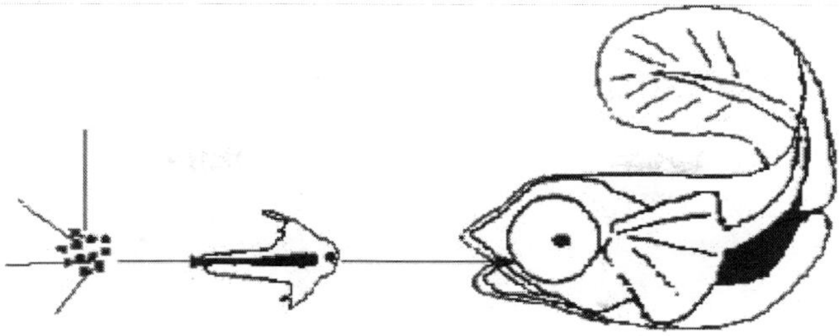

Among the live diets used in the larviculture of fish and shellfish, nauplii of the brine shrimp Artemia constitute the most widely used food item. There has been an increasing interest in culture of this crustacean after the legendery work on its utility as food for culture of larvae of aquatic organisms by Scale in 1933. Although Artemia is known to man for centuries, its use as aforesaid apparently began only in the 1930s, when several investigators found that it made an excellent food for newly hatched fish larvae. Today more than 85% of the marine organisms which are bred in aqua hatcheries world over are fed with Artemia nauplii as a sole diet or in combination with other live or inert food items. Commercially important fish and crustaceans produced in hatcheries must be fed live food during the early stages, because the larvae do not assimilate prepared diets effectively. The brine shrimps do not possess any anatomic defence mechanism against predation; consequently they are an easy prey for prawn and fish larvae. Feeding aquatic larval species with freshly hatched Artemia is like channelising mother's milk to the new borne off spring to provide important growth molecules. Foremost among these are the highly unsaturated fatty acids (HUFAs), caroteonoids and proteins. Young aquatic animals nourished with this foundation diet produce flourishing progeny, exhibit enhanced growth because of more efficient feed conversion and they develop greater resistance to diseases. After the initial work, considerable emphasis has been laid by many researchers on the importance of n-3 fatty acids for fish larvae and now various techniques have been used to improve the n-3 HUFA content in live food organisms.

What is Bioenrichment of Artemia Nauplii?

Artemia are non-selective, continuous filter feeders. Anything that is of the right particle size (10 - 50 microns) can be taken into the gut of live Artemia. Thus, by bio-encapsulating specific amounts of particulate or emulsified products rich in highly unsaturated fatty acids in the brine shrimp meta nauplii, the nutritional quality of the Artemia can be further tailored to suit the predator's requirements. Application of this method of bio-encapsulation is also called "Artemia enrichment" or "boosting".

How to Enrich Artemia Nauplii?

In Artemia, the commonly applied technique to boost the HUFAs is to provide a 12 - 24 hour enrichment period after hatching. Since Artemia nauplii moult into the second instar after 8 hours of hatching and are non-selective particle feeders, simple methods have been developed to incorporate different kinds of products into the Artemia prior to using them for feeding predatory larvae.

Steps Involved

Freshly hatched nauplii are washed with freshwater and transferred to the enrichment tank at a density of 100 - 300 nauplii / ml. It is not recommended to add the enrichment emulsion directly to the hatching water due to the high levels of bacteria that proliferate during incubation of the eggs. Addition of the enrichment emulsion with high levels of bacteria would drop the dissolved oxygen level significantly and reduce hatching efficiency. Thus, the hatched Artemia should be harvested, rinsed through fine mesh net and then transferred to the enrichment tank containing disinfected sea water maintained at 25 + 2°C.

Strong aeration should be given to maintain an optimum DO level of 4 ppm. If mortalities occur during the enrichment process, aeration should be increased so that there is continuous upwelling of water. This will not disturb feeding by nauplii. Adding sodium bicarbonate to maintain a pH above 7 also helps to improve quality.

In the normal course about 0.3 g / litre (0.5 ml / litre) is sufficient to achieve optimum enrichment. After 18 to 24 hours of hatching time, another 0.5 ml of the enrichment diet per litre is added 2 hours before harvesting. Instead of adding the drops directly to the hatching vessel, required amount of emulsion is first dissolved in a test tube filled with sea water and then added to the hatching vessel.

In commercial hatcheries enrichment is generally carried out once per day (24 hours) but minimum of 12 hours of enrichment is needed for Instar II nauplii to fully pack the gut and some of the tissues. Enriched live prey is stored at 4°C and fed at intervals to the fish larvae throughout the day. Usually prey is not immediately consumed by fish larvae but remain in the larval rearing tanks for many hours with subsequent loss of HUFA. As the fatty acids are taken up by the Artemia, their fatty acid profiles change according to the duration of the enrichment period.

Newly hatched Artemia has 7.0 mg total fatty acids/100 mg dry weight with no detectable levels of DHA. After enrichment for 12 h, the total fatty acids increase significantly to 10.3 mg / 100 mg and the Artemia has significantly higher amounts of essential fatty acids in the n-3 and n-6 families. After 24 h of enrichment, significantly higher levels of essential fatty acids and total fatty acids are achieved.

Nauplii should be transferred or exposed to the enrichment medium as soon as possible before first feeding, so that they may begin feeding immediately after the opening of the alimentary tract. As a result, the increase of nauplii size during enrichment can be minimized.

Artemia need to be observed regularly under a microscope to check lipid droplets inside the foregut. Efficient enrichment gives a reddish colouration and poor enrichment a pale colouration with no lipid droplets.

When the enriched Artemia is harvested and introduced into the larval tank, the filter feeding behaviour continues as the Artemia filter whatever particles are in tank water and "poop out" the enrichment booster from their gut. So, the sooner the fish or invertebrate larvae are able to consume the enriched Artemia, HUFAs at increased levels are delivered to the prey organism. Hence, the ad-libitum feeding regime could be very useful to achieve maximum benefit.

To avoid the process of enriching Artemia batch after batch before feeding, storage of the enriched Artemia is recommended. After one feeding, remainder Artemia is poured into a shallow dish along with some clean sea water and stored in the refrigerator. At 4 °C the metabolism is slowed down and the ingested enrichment will stay in tact for at least 24 hours for later feeding. Larger quantities of enriched nauplii can be produced but they require storage at 5 °C at a density of 5,000 number / ml with appropriate aeration. Nauplii stored in this way maintain their enrichment for upto five days.

How to Prepare Enrichment Diets?

In commercial hatcheries, where, enrichment boosters are required in bulk, simple home made techniques can be used to prepare boosters with available marine oils rather than going in for commercial expensive enrichment products.

References

www.fao.org

www.wikipedia.com

www.fishbase.org

Aquaculture - Principles and Practices / T.V.R Pillay & M.N. Kutty /Wiley India Pvt Ltd Publications.

Text Book on Fish Nutrition & Feed Technology / S.Athithan, N.Felix & M.Venkatasamy / Astral International Pvt. Ltd. Publications.

Practical Book on Fish Nutrition and Feed Technology / S.Athithan / Narendra Publishing House.

Manual on Fish Nutrition and Feed Technology / N.Felix, B.Ahilan and S.Athithan / Fisheries College & Research Institute, Thoothukudi, Tamil Nadu, India.

Fish Nutrition in Aquaculture by Sena S.Desilva and Trevor A. Anderson / Springer Publications.

Feed Management in Intensive Aquaculture by Stephen Goddard /Springer Publications.

Text Book of Aquaculture by M. Srinivasulu Reddy& K.R.S.Sambasiva Rao / Indian Books and Periodicals Publications.

Hand Book on Live Feed / MPEDA Publications.

Training Manual on phytoptankton identification / taxonomy CMFRI, Cochin Rotifer Production (as a First Feed Item) for Intensive Finfish Larviculture / Publication 600–105.

Important Live Food Organisms and Their Role in Aquaculture / 2012 / Pronob Das, Sagar C. Mandal, S. K. Bhagabati, M. S. Akhtar and S. K. Singh / Frontiers in Aquaculture, Pages 69–86 Edited by : Munilkumar Sukham.

Culturing Algae/1998/Hilary Belcher and Erica Swale / Natural Environment Research Council /UK.

Freshwater Algae: Identification and Use as Bioindicators / 2010 / Edward G. Bellinger and David C. Sigee C / Wiley & Sons, Ltd Publication.

Live Feeds in Marine Aquaculture / 2003 / Edited by Josianne G. Støttrup, Ph.D Danish Institute for Fisheries Research, Charlottenlund, Denmark and Lesley A. McEvoy, PhD North Atlantic Fisheries College, Shetland Isles, UK / Blackwell Science Publishers.

Phytoplankton Culture for Aquaculture Feed / 2010 / SRAC Publication No. 5004 .

Principles of Aquaculture by Robert R. Stickney / Wiley Publishers.

Fish Nutrition by John E. Halva / Academic Press Publications.

Encyclopedia of Aquaculture by Robert R.Stickney / Wiley-Interscience Publications.

Handbook of Fisheries and Aquaculture / ICAR Publications, New Delhi.

Index

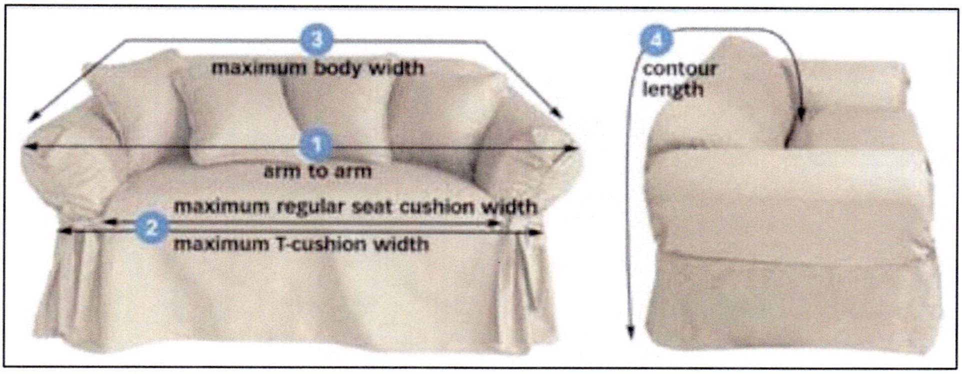

Figure 4.1: Application of Home Textiles in Living Room. (p. 27)

Figure 4.2: Application of Home Textiles in Kitchen and Dining Room. (p. 28)

Figure 4.3: Application of Home Textiles in Bed Room. (p. 29)

Figure 4.4: Application of Home Textiles in Bathroom (p. 30).

Figure 4.5: Application of Home Textiles in Pooja Ghar Materials (p. 31).

Figure 4.6: Application of Home Textiles in Children Room (p. 32).

Figure 5.2: Example of Needle Felt (p. 40).

Figure 5.8: Appearance of Cut Piles Saxony Carpet (p. 42).

Figure 5.9: Appearance of Textured Cut Pile Saxony Carpet (p. 42).

Figure 5.10: Appearance of Frieze Carpet (p. 42).

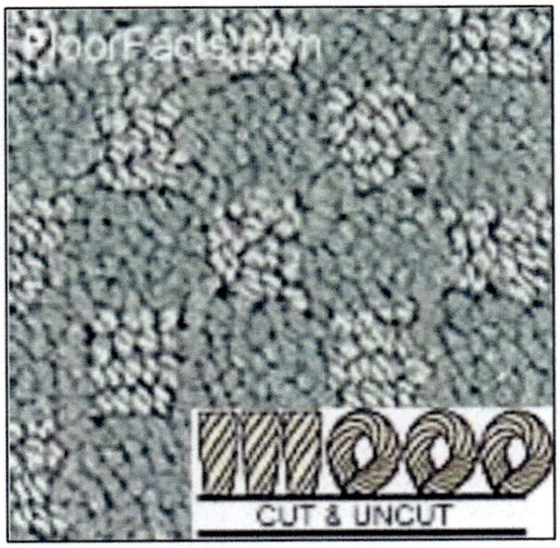

Figure 5.11: Appearance of Cut and Uncut Patterned Carpet (p. 43).

Figure 5.12a: Appearance of Multi-Level Loop Carpet (p. 43).

Figure 5.12b: Appearance of Level Loop Carpets (p. 43).

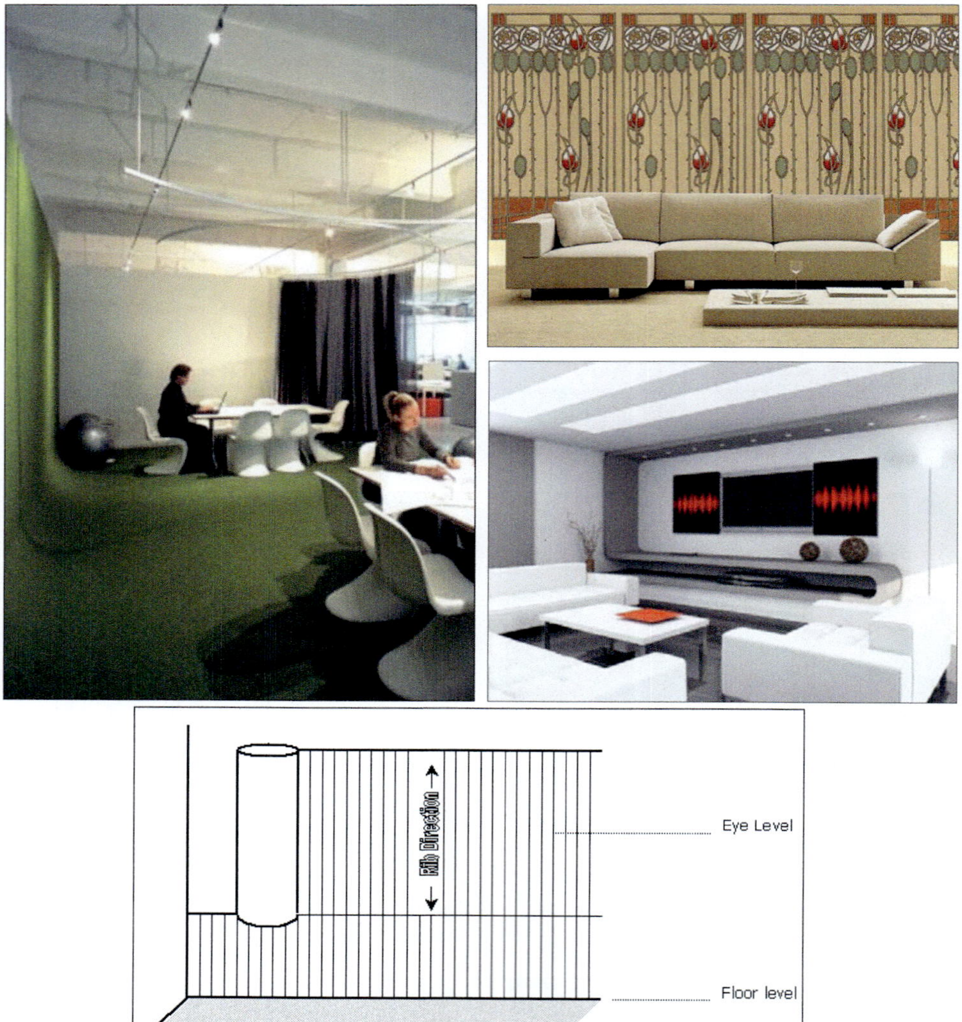

Figure 5.14: Application of Carpets as Wall Coverings (p. 45).

Figure 5.17: Double Rapier Weaving Machine for Loop Pile Carpets CLX81 (p. 48).

Figure 5.18: Weaving Machine for the Production of High Pile Carpets SRX82 (p. 49).

Figure 5.28: Construction of Earth Carpet (p. 58).

Figure 5.29: Construction of Thinking Carpets (p. 59).

Figure 6.2: Application of Textiles in Oven Mittens (p. 75).

Figure 6.3: Application of Textiles in Pot Holder (p. 75).

Oven mitt

Potholder

Kitchen Napkin

Apron

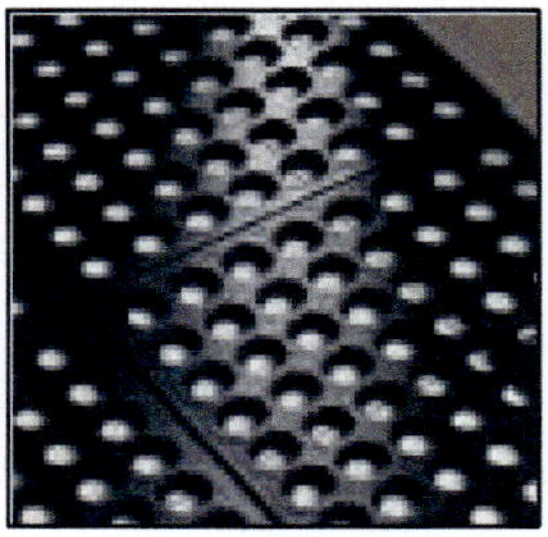

Kitchen Mats

Doilies

Dining Cloth

Kitchen Curtain

Tea Cozy

Table Runner

Kitchen Rug

Figure 6.1: Application of Textiles in Kitchen Room (p. 74).

Figure 6.4: Application of Textiles in Apron (p. 76).

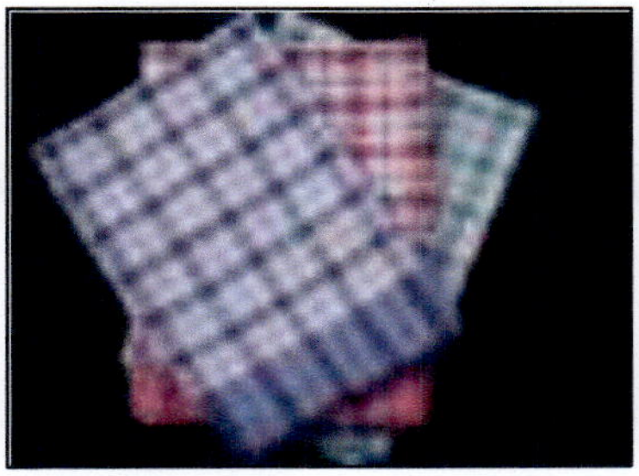

Figure 6.5: Application of Textiles in Napkins (p. 76).

Figure 6.6: Application of Textiles in Doilies (p. 76).

Figure 6.8: Application of Textiles in Dining Table Cloth (p. 77).

Figure 6.9: Application of Textiles in Tea Cozy (p. 78).

Figure 6.10: Application of Textiles in Kitchen Curtain (p. 78).

Figure 6.11: Application of Textiles in Table Runner (p. 79).

Figure 6.12: Application of Textiles in Rugs (p. 79).

Figure 7.1: Placemats (p. 83).

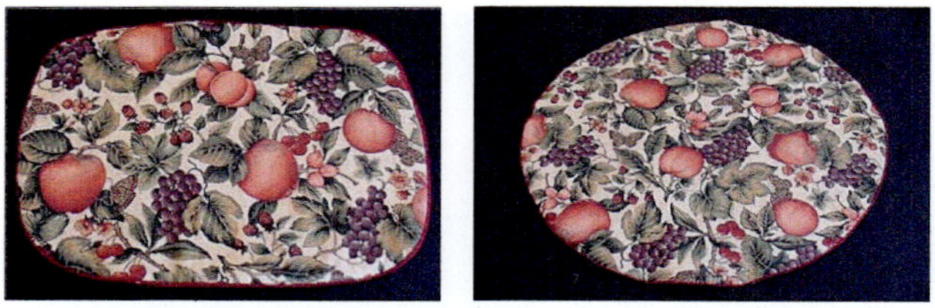

Figure 7.2: Fabric Placemats (p. 84).

Figure 7.3: Reversible Placemats (p. 85).

Figure 7.4: Quilted Placemats (p. 85).

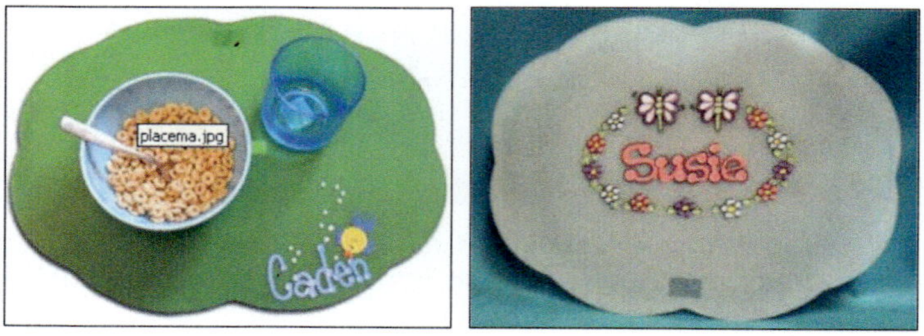

Figure 7.5: Plastic Placemats (p. 86).

Figure 7.6: Bamboo Placemats (p. 86).

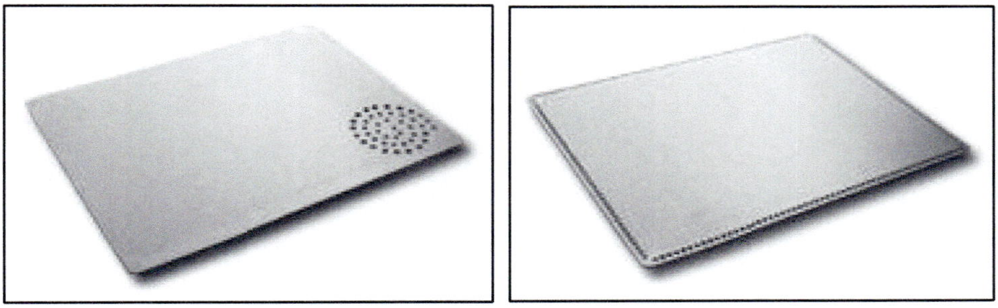

Figure 7.7: Steel Placemats (p. 86).

Figure 7.8: Stone Placemats (p. 86).

Figure 7.10: Template Cutting for Placemats (p. 88).

Figure 7.11: Cutting of Traced Lines (p. 88).

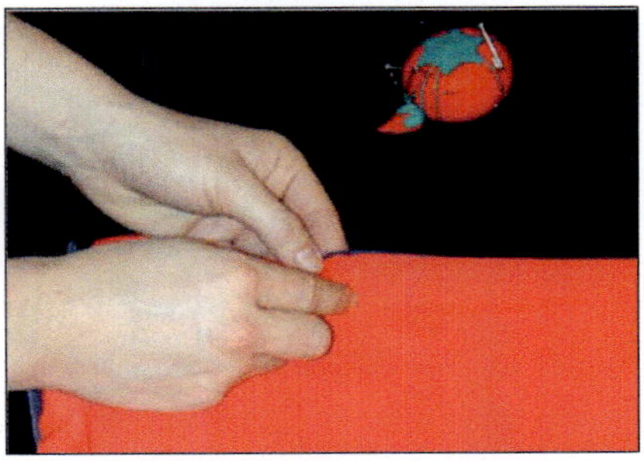

Figure 7.12: Pinning of Placemats (p. 89).

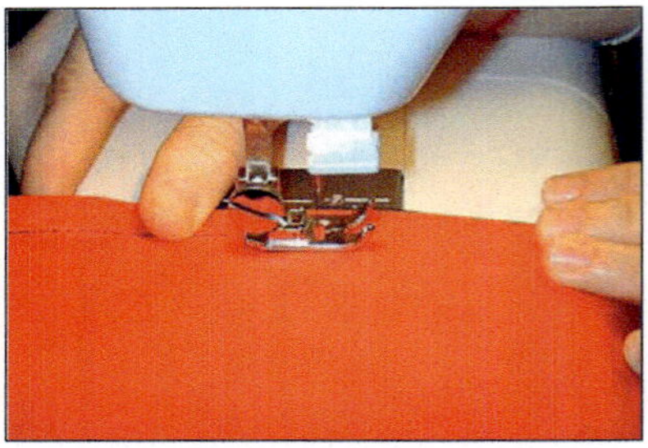

Figure 7.13: Sewing of Components for Placemats (p. 89).

Figure 7.14: Stone Placemats (p. 91).

Figure 7.16: Types of Table Cloth (p. 95).

Figure 8.1: Cottage Curtains (p. 100).

Figure 8.2 Hourglass Curtains (p. 100)

Figure 8.3: Curtains Attached with Valance (p. 101)

Figure 8.4: Draperies (p. 102).

Figure 8.6: Balloon Valances (p. 103).

Ascot Valance

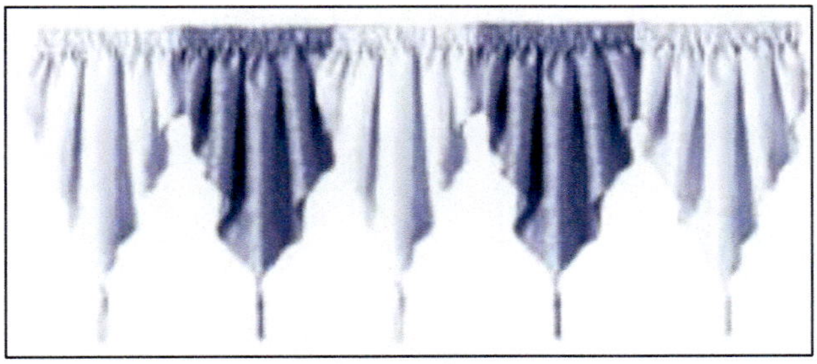

Victoria Valance

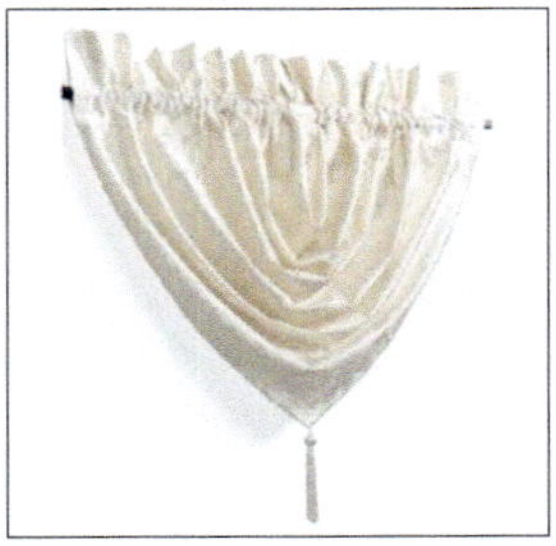

Water Fall Valance

Figure 8.8: Types of Tapered Valances (p. 104).

Figure 8.9: Construction of Shades (p. 105).

Figure 8.10: Constructions of Swags (p. 106).

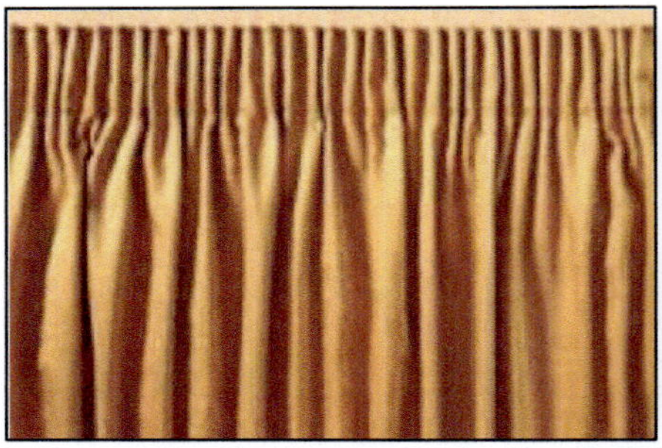

Figure 8.13: Pencil-Pleated Heading Tapes (p. 108).

Figure 8.14: French, Triple or Pinch Pleated Heading Tape (p. 108).

Figure 8.15: Tab Top Headings Tape (p. 108).

Figure 9.2: Stitching of Quilted Fabric (p. 116).

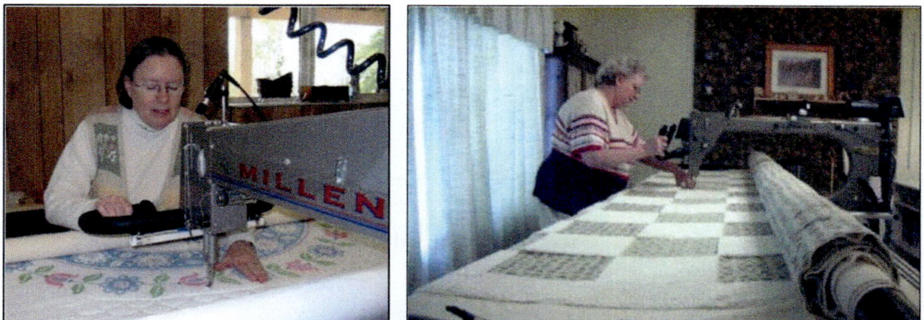

Figure 9.1: Example of Machine Quilting (p. 116).

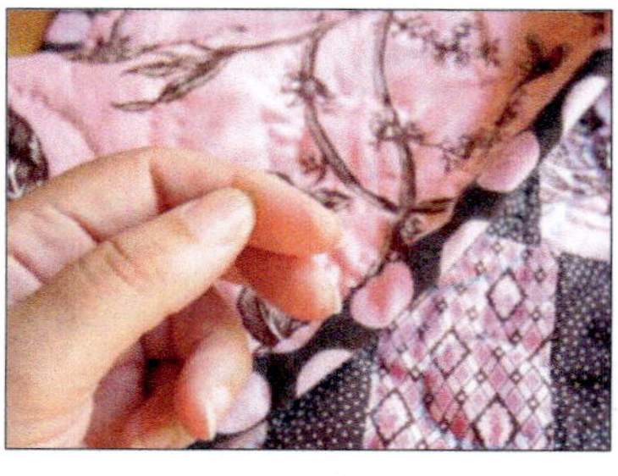

Figure 9.3: Example of Hand Quilted Fabric (p. 117).

Figure 9.4: Various Designs of Bedspread (p. 119).

Figure 9.5: Geometric Designs of Bedspread (p. 120).

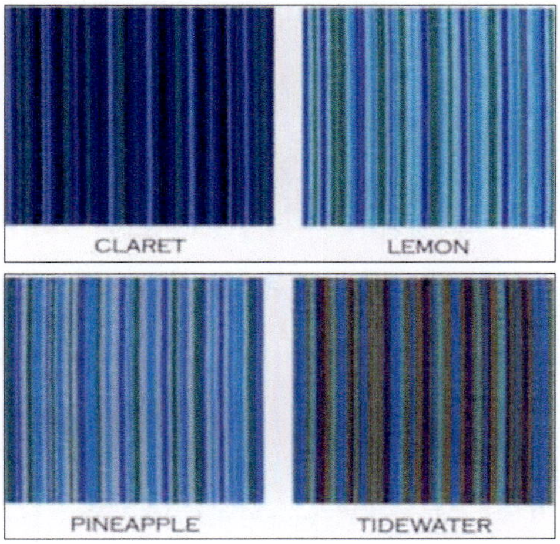

Figure 9.6: Stripe Designs of Bedspread (p. 120).

Figure 9.7: ***Chenille*** **Bedspreads (p. 121).**

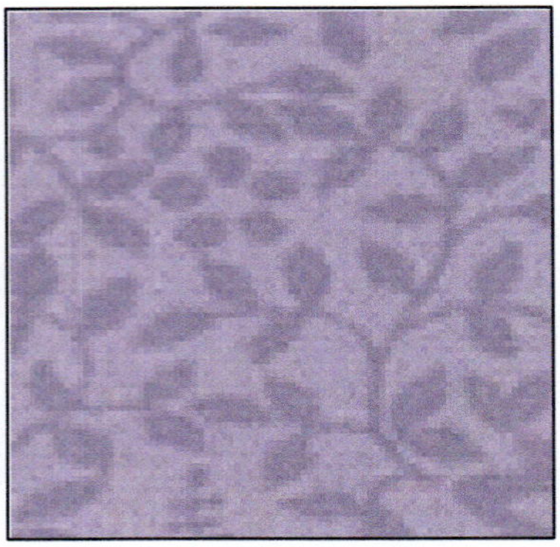

Figure 9.8: 100 per cent Organic Cotton, Floral Pattern (p. 121).

Figure 9.11: Various Designs in Pillows (p. 125)

Back Stitch

Brick Stitch

Chain Stitch

Cross-Stitch

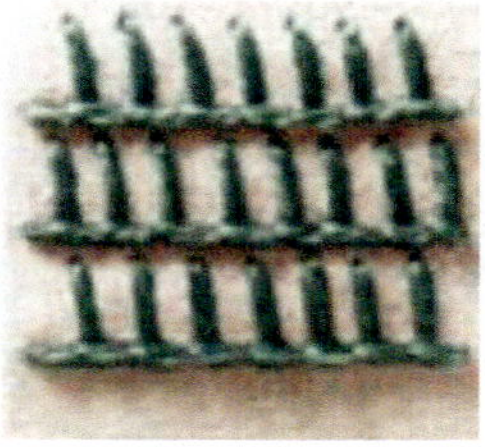

Blanket Stitch

Couching

Daisy Stitch

Coral Stitch

Fishbone Stitch

Fly Stitch

Feather Stitch

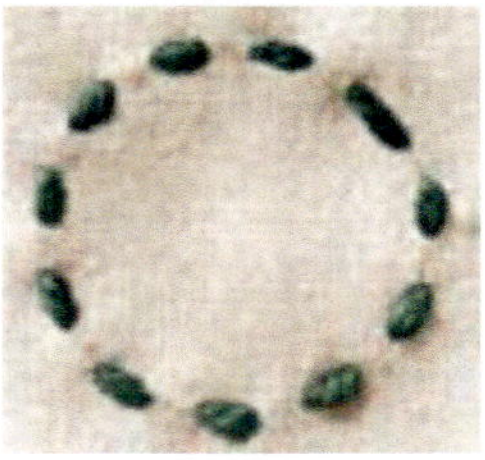

Herringbone Stitch

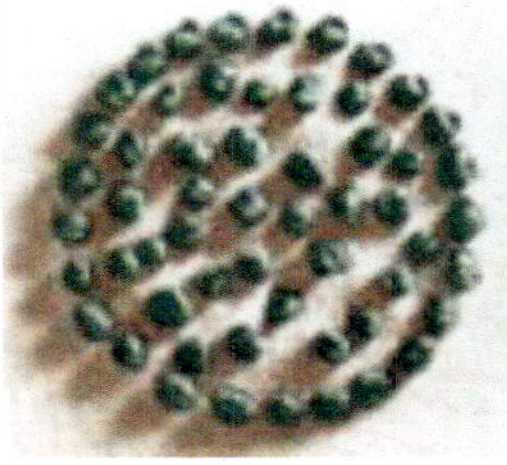

Star filling stitch
Rumanian Stitch

Overcast Stitch

French knot

Figure 10.1: Types of Embroidery (p. 129).

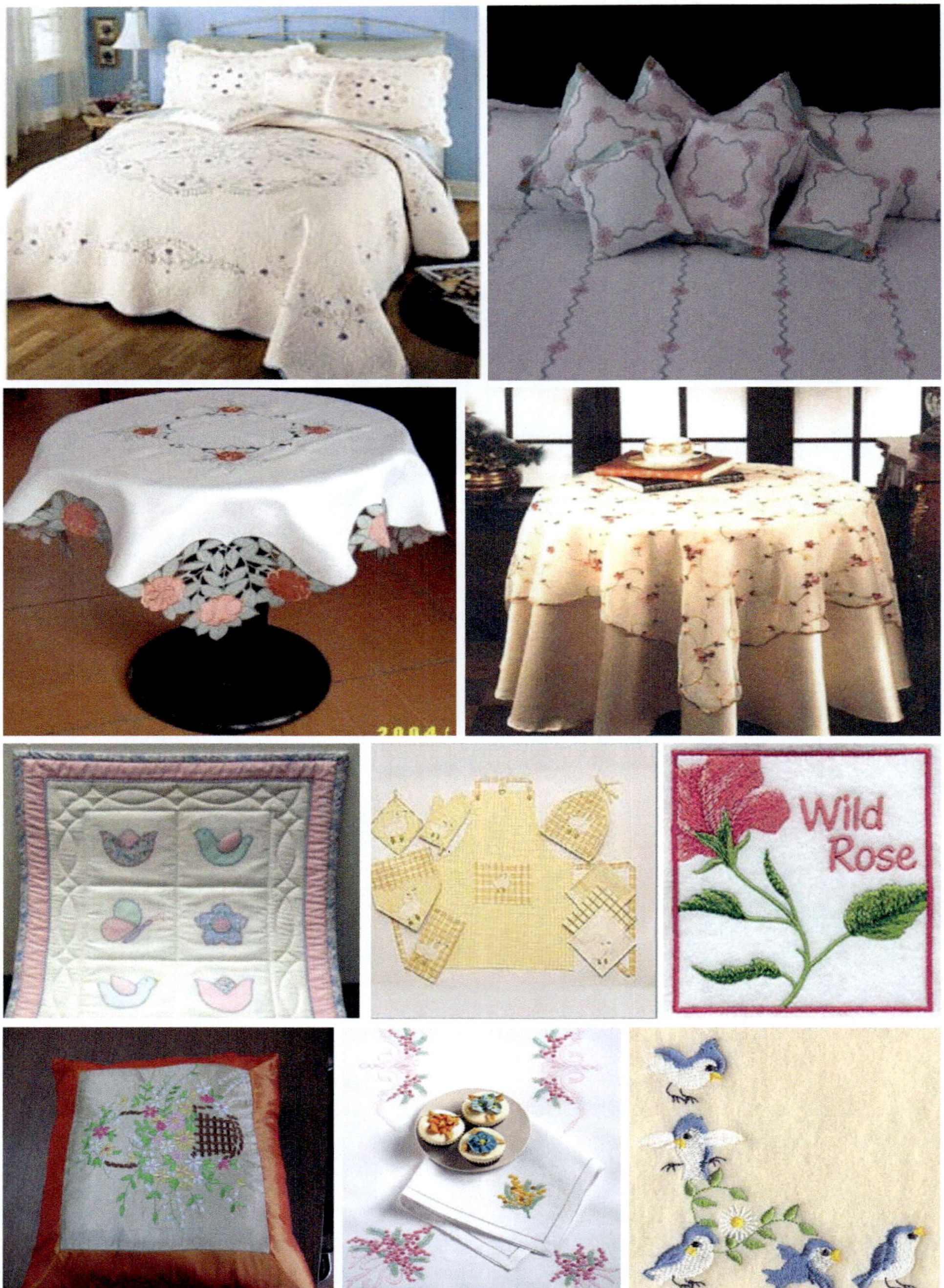

Figure 10.2: Embroidery Work in Home Textile Products (p. 132).